MAGNETIC AND ELECTRIC SUSPENSIONS

MAGNETIC AND ELECTRIC SUSPENSIONS

Types of suspensions
developed in the
Charles Stark Draper Laboratory

Richard H. Frazier
Philip J. Gilinson, Jr.
George A. Oberbeck

The MIT Press
Cambridge, Massachusetts, and London, England

This book was printed on Mohawk Neotext Offset
and bound in Columbia Millbank Vellum MBV-4660 (rust)
by Halliday Lithograph Corporation
in the United States of America.

Library of Congress Cataloging in Publication Data

Frazier, Richard Henry, 1900–
   Magnetic and electric suspensions.

   (Monographs in modern electrical technology)
   Includes bibliographical references.
   1. Magnetic suspension. 2. Electric suspension. I. Gilinson, Philip J., Jr., 1914–
joint author. II. Oberbeck, George A., 1927–   joint author. III. Title.
TK454.4.M3F7   621.319′21   73–10453
ISBN 0–262–06054–X

11 DEVELOPMENTS AND APPLICATIONS 298

FOREWORD

Mankind has come to depend strongly on understanding of the environment to help toward actions favorable for survival and continued progress. Until very recent times, information was collected by human senses, processed by human brains, and applied by human muscles in tasks to satisfy the needs and desires of men. For the long aeons of prehistory and all recorded history until about the last two centuries, human capabilities did not severely limit performance for the arrangements used to achieve real-world results. With science and engineering leading the way, technology has greatly changed this situation. In modern times, eyes, ears, touch, and other sense organs have become inadequate for gathering information; complexity and rapidity requirements of processing data have, in many cases, exceeded the capacity of the brain, while forces and torques needed to effect real-world results have gone far beyond the possibilities of muscular actions.

Power boosters coupled between command elements providing computational results and effector controls operating in the now-universal feedback pattern of servo theory have eliminated difficulties of this latter kind. Developments in electronics and communications, with emphasis on digital computers, have brought even the most complex problems of dealing with information-representing signals within the range of practical inanimate equipment of adequate capacity and acceptable size, weight, reliability, and cost. Under the pressure of strong demands from well-funded customers for arrangements to transmit, process, and use information, the technology of signal-handling subsystems has progressed rapidly in recent years and will surely continue its advance during the foreseeable future. These circumstances make certain that for any situation in which all the essential data are accurately represented by signals overall system performance will, in general, not be limited by imperfections in transmission, processing, and application of the information associated with these signals. This statement, of course, implies that the best of current technology is available for use in any equipments that may be involved.

With the world of signals internal to a system free from significant inaccuracies, the imperfect behavior of the sensor-interface "window" looking outward toward the environment on one side and coupling with computation and application on the other must be the circumstance that limits overall system performance. For this reason the direction of especially great efforts toward the improvement of sensors is essential.

In general, any sensor is affected by the input quantity it is intended to receive and produces output signals representing desired functions of this input. Effects on the output signal from quantities other than the prime input and undesired components that may be generated inside the sensor lead to inaccuracies in the output signal. Effects of this kind are propagated through the information-handling subsystems and impair the quality of overall performance.

For this reason, the improvement of sensor performance has been and remains an endeavor of primary importance for the continued progress of informetrical technology. Sensors are of many types and deal with widely varying situations in which output signal inaccuracies have their origins in a broad spectrum of effects. Some imperfections come from thermal or electro-magnetic radiation "noise" that is usually attacked by the use of low receiver temperatures and filtering techniques. Some depend upon mechanical toler-ances, eccentricities, and friction effects associated with contacts between solid parts. Among these and other possibilities, the reduction of inaccuracy-producing defects is especially important for sensors based on interactions between rigid body masses with imposed accelerations and gravitational fields. These interactions depend on Newton's laws of dynamics, which are true with substantial perfection for the range of masses, accelerations, and fields encountered in the operation of practical systems. The input quantities are usually changes in linear velocity and gravitational fields for devices that are generally called "accelerometers." Another class of instruments designed to receive angular velocities is based on gyroscopic principles.

All sensors of these types depend upon accurate restraint by some cali-brated effect against the forces generated by a gyroscopic rotor or an un-balanced mass. Widely used designs depend upon a cylindrical or spherical member containing a spinning rotor or a mass of this kind. Newton's law reactions to rotational motion or translational accelerations and gravity are ideal; the real-world problem is to support receiving elements by means that do not introduce inaccuracy-producing torques or forces. Even the very smallest rubbing friction effects are unacceptable, but pure viscous friction giving resistance to motion that is accurately proportional to relative motion between active parts and their enclosing cases can be applied as a useful factor in sensor operation. This fact makes possible the use of flotation in a viscous fluid as the means for effectively eliminating all contact between solid parts.

This principle was introduced some twenty-five years ago and has proved useful in many thousands of working instruments. Even though careful flotation with good temperature control greatly reduces erratic friction components while supplying the velocity-proportional restraints needed for integration, imperfections in balance and temperature are always present that prevent the achievement of highest sensor performance. The residual force on the moving member required to eliminate all rubbing friction effects may be only a small fraction of the total weight of the member but is absolutely necessary for consistent results of the highest quality.

The authors of this book have had the boldness to attack the problem of eliminating the last traces of rubbing friction in floated instruments by providing suspensions—very weak in terms of the total unfloated weight of the parts involved—but nevertheless basically essential for the achievement of substantially perfect results from sensors based on forces and torques produced by Newton's law reaction forces and gravitational fields. Fundamental principles and design details are the proper subject matter for the chapters that follow. Forces may be small, but performance improvements span orders of magnitude in providing advancing technology with the capabilities required by the progress of our modern societies. Readers should find the text coming from over two decades of imaginative conceptions, excellent engineering, and effective technology not only informative but also interesting as a saga of perseverance and achievement.

Charles Stark Draper

PREFACE

This monograph is primarily an account of the development of magnetic and electric suspensions at the Charles Stark Draper Laboratory.* The development at the Draper Laboratory has related mostly to the suspension of elements in the guidance systems for airplanes, submarines, missiles, and space vehicles, though the devices have numerous other applications, some of which are described in Chapter 11. Other types of suspensions have been developed elsewhere for various purposes over a considerable period of years. References 1 through 35, beginning on page 349, are intended to give a broad overview of the Draper Laboratory development up to recent years and to give a few leads to the development of other types of suspensions, especially suspensions for rapidly rotating bodies, for which the Draper Laboratory suspensions are not in general intended, though such adaptations are being developed.

A general bibliography has not been attempted, because Reference 35 considerably serves that purpose, to 1964. Beyond Reference 36, the list relates almost entirely to Draper Laboratory devices. Recent developments, especially for active suspensions and new applications, are given in Chapters 9 and 11, and much previously unpublished material is given throughout the text. Readers interested in the general investigation of the field can make a good start from Reference 35 and from the bibliography in Reference 24. Interested readers, of course, will find that one reference usually leads to others, but probably much interesting information is still buried in the files of manufacturers, laboratories, and other agencies where work has been done.

The authors are indebted to the Draper Laboratory for the use of many illustrations and much text material from its files and reports, and to the personnel who developed some of the material, whose names appear at appropriate places. Since the galleys for this book were set, the Draper Laboratory has been divested from the Massachusetts Institute of Technology (July 1, 1973), substantially reorganized, and somewhat reorientated; now it is an independent nonprofit corporation, though certain mutually advantageous ties have been retained. In the credit lines given to various staff members in reference to their work, the titles used are those held prior to divestation. Thanks are due to all these people and to numerous other

* Formerly the Instrumentation Laboratory, Massachusetts Institute of Technology.

colleagues for direct or indirect assistance. Most of all, the authors wish to express their appreciation for the many years of inspiration and leadership of the laboratory by Charles Stark Draper.

Richard H. Frazier
Philip J. Gilinson, Jr.
George A. Oberbeck

Cambridge, Massachusetts

MAGNETIC AND ELECTRIC SUSPENSIONS

1 ORIENTATIONAL SURVEY

Here the various types of electromagnetic suspensions are cited, and the scope and basis of the treatment in this monograph are indicated.

1-1 Active and Passive Suspensions

Magnetic and electric suspensions act to maintain the position of a body along or about one or more axes attached to a certain reference frame, through the force actions of magnetic or electric fields. Such suspensions generally are statically unstable. Stable action can be achieved either by sensing the position of the body continuously or at intervals to enable the force fields to be controlled by servo action with sufficient rapidity to prevent the body from departing from its desired position by more than a tolerable amount or by adjustment of the parameters of the energizing circuits themselves. The servo-controlled suspension may be called an *active* type, whereas a suspension that is inherently stable owing to adjustment of the parameters of its energizing circuits may be called a *passive* type.

For an active suspension, position may be sensed by a variety of electrical or mechanical probes, by change of electrical or magnetic parameters, possibly utilizing bridge networks, by light beams, or by other methods. Servo-controls may be superimposed on a passive suspension to increase its stiffness or because it is stable only over a certain range of departure from a desired position. Permanent-magnet suspensions are stable in repulsion at least along one axis and hence are passive, but they are not considered in this monograph. Either type of suspension may be subject to oscillation, which may be controlled by servo design or by mechanical or electromagnetic damping and parameter adjustment, depending on the situation.

1-2 The Draper Laboratory Development

Both active and passive suspensions have many forms and applications. As indicated in the Preface, the suspensions developed in the Draper Laboratory are intended primarily to keep elements in the guidance systems of airplanes, submarines, missiles, and space vehicles essentially fixed with respect to a designated reference without friction, in the face of considerable motion of the vehicle, involving high acceleration, vibration, or shock. Hence economy of space, weight, and energy and reliability and excellence of performance are paramount. The suspended element usually is floated in a suitable fluid, which serves also for damping, and thus the suspension problem is considerably eased. If the element needs to rotate continuously, the use of such a fluid generally is not practical, and the suspension problem is more difficult.[16,17]* The object is not to sustain weight but to hold the floated member centered away from its mechanical bearings, against the small decentering end- and side-loading forces, inherent in other associated magnetic or electric components or that arise from temperature gradients, and the large inertia forces that arise from disturbances in the motion of the vehicle. The small decentering forces tend to cause drift in the quiescent state, whereas the large inertia forces due to vehicle motion tend to drive the floated member to its mechanical bearings.

The development started with an active type of suspension for which the electronics and circuitry for sensing and servo action required considerable bulk and weight.[1,2,3,7,8] The original scheme was succeeded by a rather ingenious passive suspension of great simplicity and space economy in which self-stabilization was achieved merely by appropriate tuning of the circuitry of the suspension itself but with some sacrifice of force and stiffness.[4-8] More recently, with the perfecting of transistors, other solid-state devices, and integrated circuits, making possible great reduction in bulk and weight of sensing and servo circuitry, design has returned to the active suspension for the exploitation of the possibilities of increased force and stiffness without the accompanying cost in bulk and weight of auxiliary apparatus. This development, which has made possible also substantial energy economy through pulsed operation, time sharing with other functions or other components, and elimination or reduction of continuous quiescent currents, is the subject of Chapter 9. The first stage of this active suspension development is largely ignored, except for background purposes, because the apparatus is obsolete, and the emphasis is placed on the currently important passive and active types of suspensions as developed by the laboratory. The earlier active suspensions are briefly described as introduction to the development of the active suspensions currently in use.

1-3 Basis of Force Computations[36,37]
The gaps in the suspensions developed by the Draper Laboratory are very short compared with the cross-sectional dimensions, being usually a few mils, so that fringing effects generally are negligible within the accuracy with which

---

*Superior numbers are for the references listed beginning on page 349.

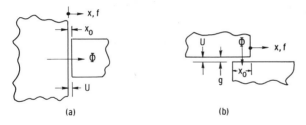

Figure 1-1 Magnetic pull between gap faces: (a) in the flux direction, (b) normal to the flux direction.

computations can predict performance. Basic analysis then is quite simple compared with analysis of long-gap configurations[20,21] that require the use of field theory rather than circuit theory, even though conditions may be regarded as quasi-static. Though the gap surfaces may be cylindrical or conical, they may be treated as parallel flat-faced gaps, with motion either parallel to the direction of the flux lines, Figure 1-1(a), or normal to the direction of the flux lines, Figure 1-1(b), with good approximation.

For Figure 1-1(a), the magnetic pull is

$$f = -\frac{\Phi^2}{2}\frac{d\mathcal{R}}{dx} = \frac{\Phi^2}{2\mu A} = \frac{U^2}{2}\frac{d\mathcal{P}}{dx} = \frac{U^2}{2}\frac{\mu A}{(x_0 - x)^2} = \frac{\mu \mathcal{H}^2 A}{2} = \frac{\mathcal{B}^2 A}{2\mu}, \quad (1\text{-}1)$$

and for Figure 1-1(b),

$$f = \frac{U^2}{2}\frac{d\mathcal{P}}{dx} = \frac{U^2}{2}\frac{\mu w}{g} = \frac{\mu \mathcal{H}^2 wg}{2} = -\frac{\Phi^2}{2}\frac{d\mathcal{R}}{dx} = \frac{\Phi^2}{2}\frac{g}{\mu w(x_0 + x)^2} = \frac{\mathcal{B}^2 wg}{2\mu}. \quad (1\text{-}2)$$

Here
Φ = gap flux,
$\mathcal{R}$ = gap reluctance,
$\mathcal{P}$ = gap permeance = $1/\mathcal{R}$,
μ = gap permeability,
A = cross-sectional area of gap, Figure 1-1(a),
w = width of gap, Figure 1-1(b),
U = magnetic drop across gap,
$\mathcal{H}$ = magnetic field intensity in gap,
and
$\mathcal{B}$ = magnetic flux density in gap.

If corresponding gaps are traversed by electric fields, the forces are

$$f = -\frac{Q^2}{2}\frac{dS}{dx} = \frac{Q^2}{2\varepsilon A} = \frac{V^2}{2}\frac{dC}{dx} = \frac{V^2}{2}\frac{\varepsilon A}{(x_0 - x)^2} = \frac{\varepsilon \mathscr{E}^2 A}{2} = \frac{\mathscr{D}^2 A}{2\varepsilon} \qquad (1\text{-}3)$$

and

$$f = \frac{V^2}{2}\frac{dC}{dx} = \frac{V^2}{2}\frac{\varepsilon w}{g} = \frac{\varepsilon \mathscr{E}^2 wg}{2} = -\frac{Q^2}{2}\frac{dS}{dx} = \frac{Q^2}{2}\frac{g}{\varepsilon w(x_0 + x)^2} = \frac{\mathscr{D}^2 wg}{2\varepsilon}, \qquad (1\text{-}4)$$

in which
Q = electric charge on gap surface,
S = gap elastance = $1/C$,
C = gap capacitance,
ε = capacitivity of gap,
V = voltage drop across gap,
$\mathscr{E}$ = electric field intensity in gap,
and
$\mathscr{D}$ = electric flux density (displacement) in gap.

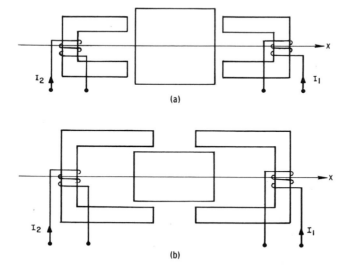

(a)

(b)

Figure 1-2 Schematic for single-axis magnetic suspension: (a) with the axis in the flux direction, (b) with the axis normal to the flux direction.

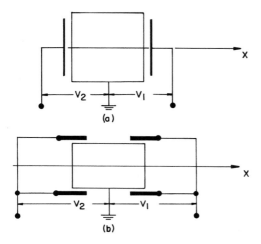

Figure 1-3 Schematic for single-axis electric suspension analogous to Figure 1-2, suspended member grounded.

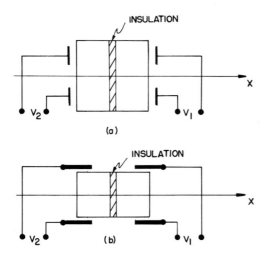

Figure 1-4 Schematic for single-axis electric suspension analogous to Figure 1-2, suspended member ungrounded.

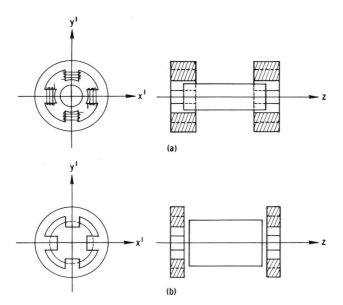

Figure 1-5 Schematic for three-axis magnetic suspension.

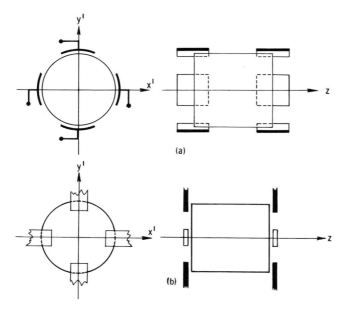

Figure 1-6 Schematic for three-axis electric suspension.

Either gap configuration can be used for the suspension of a body along one axis, as illustrated in Figures 1-2 through 1-4 or along two or three axes, as illustrated in Figures 1-5 and 1-6. If the stators of Figures 1-5(a) and 1-6(a) do not overhang the rotor, no centering action is achieved along the longitudinal axis except by fringing fluxes (if the ends of the rotor are near the outsides of the stators), and the suspension is effective only radially. The action of the Figure 1-1(a) configuration tends to be quite stiff but over small displacements, while the action of the Figure 1-1(b) configuration tends to be quite soft but over relatively large displacements. Therefore the Figure 1-5(a) and 1-6(a) suspension actions tend to be relatively strong radially but relatively weak longitudinally, whereas the situation for the Figure 1-5(b) and 1-6(b) suspensions is vice versa. However, the Figure 1-1(b) radial action tends to be impractically weak and therefore is little used, so that a compromise may be made by tapering the gaps to make them conical, as considered in Section 7-6, or separate electromagnets may be used for longitudinal and radial control, as mentioned in Sections 7-7 and 11-1.

Owing to fluid flotation and the highly overdamped condition of the mechanical system, the motion of the suspended body is so slow that it does not appreciably affect the computation of the centering forces. The basic equations, (1-1) through (1-4), are quite simple in terms of the indicated quantities for the two gap configurations, and the gross computations for forces are quite simple under the assumption that materials involved are ideal. But the many complicating effects of the practical behavior of magnetic, conducting, and insulating materials that occur under a variety of operating conditions that influence the gap quantities cannot be ignored in the face of gap lengths of a few mils and performance requirements expressed in terms of a few parts in a million. Hence, though the gross computations may be rather simple, for ultimate refinement within the limitations of the art, the design and construction problems can be very severe indeed, because many phenomena that are of secondary or negligible effect in conventional design and application may become primary for highly accurate guidance and navigation. A considerable discussion of the influence of materials is in Chapter 8.

1-4 Influence of Circuitry

Examination of Figures 1-2 and 1-3 shows that the operation of the corresponding devices as suspensions depends on how they are energized and on the design of the associated circuitry.

If in Figure 1-2(a) the coils are energized from constant and equal direct current or voltage sources, the suspended member would be in unstable equilibrium along the x axis when centered and would clap to either side if disturbed. If the gap fluxes could be kept equal and fixed, under ideal conditions the member would simply float under the influence of zero net centering force. To achieve stable operation, the flux must be increased at the gaps that are increasing in length and should be decreased at the other gaps, to the extent that makes the net force oppose the displacement. If in Figure 1-2(b) the coils are energized from constant and equal direct voltage sources, the suspended member would simply float along the x axis under the influence of zero net centering force. To achieve stable operation, the magnetic drop across the gaps where the overlap is decreasing must be increased and be decreased at the other gaps, to the extent that makes the net force oppose the displacement. Incidentally, in this instance maintenance of fixed and equal gap fluxes would give stable operation.

The same arguments apply to the electric suspension of Figure 1-3, for which electric charge Q is analogous to magnetic flux Φ, and voltage V is analogous to magnetic drop U, provided that insulation is perfect. In this illustration the suspended body is grounded, a situation that can be objectionable owing to the small forces and torques introduced by grounding leads. The grounding can be avoided by use of capacitors in series, as illustrated by Figure 1-4. However, unless the two ends of the suspended member are substantially insulated from each other, the two circuits are coupled and therefore do not act independently, and analysis must take into account the coupling. In the absence of the dividing insulation, the suspended member would assume some potential depending on its position and the voltages across the pairs of capacitors; if the two parts are separated, they would have different potentials. These conditions of uncertain and changeable potentials may be undesirable also. Actually, for steady conditions, the potentials and forces could be considerably determined by leakage resistances. Though the electric suspension itself offers the possibility of considerable saving of weight and energy, the net savings depend on the associated circuitry and apparatus that may be required. The electric suspension has been little used primarily because very large electric gradients are required to produce forces comparable with the forces of the magnetic suspension in the same frame sizes.

The problem of stable suspension therefore comes down to the control of magnetic Φ or U, or electric Q or V of the gaps, with respect to displacement of the suspended member from a neutral position. For direct current or voltage supply, the suspension must be active, that is, its position must be sensed and the currents or voltages must be adjusted accordingly, by change of level, by pulsing of some kind, or by a combination of such changes.

For alternating current or voltage supply, constant-amplitude voltage for the magnetic suspensions of Figure 1-2 would give constant-amplitude Φ, in the absence of coil resistance and leakage flux, so that the suspension of Figure 1-2(a) would tend to have zero net centering force, and the suspension of Figure 1-2(b) would tend to be stable. Likewise constant-amplitude current would give constant-amplitude U, and the suspension of Figure 1-2(b) would tend to have zero net centering force, and the suspension of Figure 1-2(a) would be quite unstable. These axial and radial destabilizing forces (whether arising from direct or alternating fields) are the same as the undesirable end thrusts and side thrusts that occur in ordinary motors and generators due to imperfect centering or other dissymmetries and cause excess bearing wear, vibration, and noise. Similar stability arguments apply to the electric suspensions of Figures 1-3 and 1-4; for alternating currents the leakage resistances should have relatively small effect, but stray capacitances may become quite troublesome, depending on the frequency.

These changes in force-displacement characteristics that depend on whether the alternating supply is essentially constant-current or constant-voltage amplitude and depend on the changes in reactances with displacement of the suspended member lead to the idea of achieving stability through circuit design, in particular through tuning, which is the basis of the passive type of suspension. With alternating excitation, a suspension is not required to be passive but may be fully active in operation, or it may be hybrid.

For either the active or the passive type of suspension, the opportunities are numerous for the exercise of ingenuity in circuit design and operation. Here the entire realm of circuit theory and feedback control theory is available. Choice of equivalent voltage- or current-source operation; use of series, parallel, or series-parallel circuit arrangements, or bridge circuits, each with miscellaneous variations; use of multiplexing or time sharing, so that the same circuitry can serve for both signal- and force-producing purposes or for the operation of different devices; all have their advantages and disadvantages and are explored in following chapters. In general, for simplicity, analysis can be made on a single-axis basis without much loss in significance when translated to multiaxis operation, for small displacements. For the three-axis suspensions of Figures 1-5 and 1-6, and for structures with larger numbers of poles or plates, various couplings exist that must be taken into account for general solutions that go beyond small displacement. Algebraic complexities require that such solutions[15] be made by computer. These couplings can be avoided by isolation of pairs of magnetic poles or pairs of capacitor plates, but such isolation may be difficult for structural reasons. Multiaxis and multipolar structures are considered in Chapter 7.

1-5 The Design Problem

Technically the design procedure involves the solution of problems in magnetic and electric circuitry as related to the mechanics of the suspended member, and the favorable selection of materials, with pressure on the economy of space, weight, and energy, emphasis being always on high performance and reliability. While exotic and costly materials may be involved, that fact does not mean that cost is no object. For the quantities required, materials costs generally are minor in comparison with the engineering costs of the ingenious designs and the costs of delicate workmanship required to produce the high reliability and excellence of performance demanded. Costs therefore must be judged in the light of the consequences of failure and the importance of the mission to be accomplished.

1-6 Summary

The Draper Laboratory has developed both active and passive suspensions, primarily for use in navigation and guidance systems. Emphasis therefore is on economy of weight, space, and energy, and on excellence and reliability of performance. Such objectives are achieved through ingenuity of magnetic and electric-circuit design, careful selection and handling of materials, and excellence of workmanship.

2 IDEAL SINGLE-AXIS PASSIVE MAGNETIC SUSPENSION

To avoid the beclouding of basic ideas by immediate introduction of the complications that accompany suspension with respect to all directions and the influence of imperfections of materials and workmanship, the problem initially is limited to suspension along only one axis under somewhat idealized conditions. The principal features of the passive suspension, which in fact carry over to the more complicated situations, thus are emphasized as an aid to the understanding of the more sophisticated multiaxis designs and more complex applications.

2-1 Principle of Operation

For simplicity of illustration, the manner of operation of the self-stabilizing suspension is explained from the action of a block of magnetic material constrained to be free to move along the x axis only, as shown in Figure 2-1, under the influence of two electromagnets.* The magnetic material is assumed to be lossless, so that it has no influence on effective resistance, and to have infinite permeability, so that the magnetic circuits are not coupled and can be solved independently. At the relatively low frequency of operation of early self-stabilizing suspensions, around 400 hertz, the influence of magnetic materials can be made unimportant by use of high-permeability, low-loss, thinly laminated ferromagnetic material or by use of ferrites, but for higher

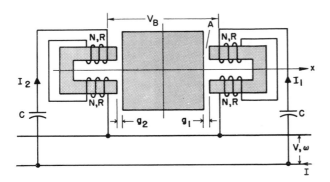

Figure 2-1 Schematic for single-axis magnetic suspension, Case (1), or Case (2) (Chapter 3).

* Reference 15, pp. 9–17.

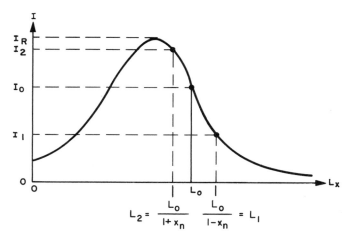

Figure 2-2 Coil current versus self-inductance, Case (1).

frequencies currently in use, around 12,800 hertz, or for pulsed operation, the effects of eddy currents, hysteresis, and disaccommodation are important. These effects, and the limitations of magnetic saturation, are considered in subsequent chapters.

If the coils in Figure 2-1 were simply excited by equal direct currents or by equal alternating currents or voltages, the block when centered would be in unstable equilibrium and would tend to clap against one pole pair or the other at the slightest disturbance. But if each coil pair is tuned with a capacitor as shown in Figure 2-1, so that the current follows the resonance curve of Figure 2-2 as coil inductance changes with change of gap length, the system can be self-stabilizing. When the block is displaced slightly, the current in the coils on the side having the shorter gaps decreases and the current in the coils on the side having the longer gaps increases; and if the changes in currents are sufficient to establish more flux in the long gaps than in the short gaps, the net force tends to recenter the block. The conditions required to achieve this self-stabilization and the equation of restoring force now are derived for the circuit connections shown in Figure 2-1; conditions for other circuit connections are analyzed in following chapters.

2-2 Force Equations, Voltage Source, Series Tuning, Case (1)
The connections used in Figure 2-1 may be described as series tuning with tuned circuits in parallel across a voltage source. This condition is designated

as Case (1). The impedances of the circuits at the respective ends are

$$Z_1 = 2R + 2j\omega(L_\ell + L_1) + \frac{1}{j\omega C} \tag{2-1}$$

and

$$Z_2 = 2R + 2j\omega(L_\ell + L_2) + \frac{1}{j\omega C}, \tag{2-2}$$

in which

$\quad R$ = resistance/coil,

$\quad 2L_\ell$ = leakage inductance/coil pair,

$$\left.\begin{aligned} 2L_1 &= \frac{(2N)^2\mu A}{2g_1} = \frac{2N^2\mu A}{g_0 - x} \\[2mm] 2L_2 &= \frac{(2N)^2\mu A}{2g_2} = \frac{2N^2\mu A}{g_0 + x} \end{aligned}\right\} \begin{array}{l} \text{self-inductances} \\ \text{from gap fluxes,} \end{array}$$

$\quad\quad N$ = turns/coil,

$\quad\quad \mu$ = permeability of gap region,

$\quad\quad A$ = cross-sectional area/gap,

g_1, g_2 = gap lengths,

$\quad\quad g_0$ = gap lengths when block is centered,

$\quad\quad x$ = displacement of block from central position,

$\quad\quad C$ = tuning capacitance,

$\quad\quad \omega = 2\pi f$ = angular frequency of source,

and

$\quad\quad f$ = frequency of source.

Here R is taken as winding resistance, but in subsequent analysis where an effective resistance R_e is used to account for losses that occur elsewhere R_w is used to designate that part of the effective resistance that is in the winding. The self-inductances L_1 and L_2 are computed without allowance for fringing, on the assumption that gap lengths g_1 and g_2 are small compared with the dimensions of cross-sectional areas A, and leakage inductances L_ℓ are assumed to remain constant. The inductances $L_\ell + L_1$ and $L_\ell + L_2$ are not true self-inductances of individual coils but include the mutual inductances of the coil pairs; in other words, they are merely half the self-inductances of the respective coil pairs. The respective currents are

$$i_1 = I_{1m} \cos(\omega t - \theta_1) \tag{2-3}$$

and

$$i_2 = I_{2m} \cos(\omega t - \theta_2), \tag{2-4}$$

based on

$$v = V_m \cos \omega t, \tag{2-5}$$

in which

I_{1m}, I_{2m} = amplitudes of respective currents,

$$\theta_1 = \tan^{-1} \frac{2\omega(L_\ell + L_1) - 1/\omega C}{2R}$$

$$\theta_2 = \tan^{-1} \frac{2\omega(L_\ell + L_2) - 1/\omega C}{2R},$$

$\quad\quad v$ = source voltage as function of time t,

and

$\quad\quad V_m$ = amplitude of source voltage.

The instantaneous restoring force acting on the block is

$$f_r = f_1 - f_2 = \frac{2\phi_1^2}{2\mu A} - \frac{2\phi_2^2}{2\mu A} = N^2 \mu A \left[\frac{i_1^2}{(g_0 - x)^2} - \frac{i_2^2}{(g_0 + x)^2} \right]$$

$$= \frac{N^2 \mu A}{g_0^2} \frac{V_m^2}{8R^2} \left\{ \frac{1 + \cos 2(\omega t - \theta_1)}{[Q_0 - (Q_0 - Q)(1 - x_n)]^2 + (1 - x_n)^2} \right.$$

$$\left. - \frac{1 + \cos 2(\omega t - \theta_2)}{[Q_0 - (Q_0 - Q)(1 + x_n)]^2 + (1 + x_n)^2} \right\}, \tag{2-6}$$

in which

ϕ_1, ϕ_2 = instantaneous fluxes/pole at respective ends,

$$Q_0 = \frac{2\omega L_0}{2R} = \frac{\omega L_0}{R},$$

$$2L_0 = \frac{2N^2 \mu A}{g_0},$$

$$Q = \frac{2\omega L - 1/\omega C}{2R} = \frac{\omega L}{R} - \frac{1}{2\omega CR} = Q_L - Q_c,$$

$$L = L_\ell + L_0,$$

and

$$x_n = \frac{x}{g_0} = \text{normalized displacement}.$$

This instantaneous restoring force has a steady or average component super-posed on which is a double-frequency oscillating component. On the assumption that the source frequency is sufficiently high that at double source frequency the motion of the block is inappreciable under the influence of the oscillating component of force, owing to the mass and the damping of the block, the relation between force and steady displacement is based on the average force:

$$F_{Vs} = \frac{N^2 V^2 \mu A}{4R^2 g_0^2} \left\{ \frac{1}{[Q_0 - (Q_0 - Q)(1 - x_n)]^2 + (1 - x_n)^2} \right.$$

$$\left. - \frac{1}{[Q_0 - (Q_0 - Q)(1 + x_n)]^2 + (1 + x_n)^2} \right\}$$

$$= \frac{N^2 V^2 \mu A}{4R^2 g_0^2} \left\{ \frac{4(Q^2 - Q_0 Q + 1)x_n}{\{[Q_0 - (Q_0 - Q)(1 - x_n)]^2 + (1 - x_n)^2\}}{\times \{[Q_0 - (Q_0 - Q)(1 + x_n)]^2 + (1 + x_n)^2\}} \right\}, \qquad (2\text{-}7)$$

in which V is the rms value of v. The force can be written in normalized form,

$$F_n = \frac{F_{Vs}}{F_V} = \frac{4(Q^2 - Q_0 Q + 1)x_n}{\{[Q_0 - (Q_0 - Q)(1 - x_n)]^2 + (1 - x_n)^2\}}{\times \{[Q_0 - (Q_0 - Q)(1 + x_n)]^2 + (1 + x_n)^2\}}, \qquad (2\text{-}8)$$

in which the base for normalizing,

$$F_V = \frac{N^2 \mu A}{g_0^2} \frac{V^2}{4R^2},$$

is the pull on one side of the block when the current is at resonance with the block centered. Depending on convenience in computing, alternate forms of terms in the denominators of Equations (2-6), (2-7), and (2-8) may be written as

$$Q_0 - (Q_0 - Q)(1 - x_n) = Q_0 x_n + Q(1 - x_n) = Q + (Q_0 - Q)x_n$$

and

$$Q_0 - (Q_0 - Q)(1 + x_n) = -Q_0 x_n + Q(1 + x_n) = Q - (Q_0 - Q)x_n.$$

2-3 Currents and Flux Densities

Equation (2-7) for average force can be obtained by putting rms currents

$$I_1 = \frac{V}{Z_1} = \frac{V}{2R} \frac{1 - x_n}{\sqrt{[Q + (Q_0 - Q)x_n]^2 + (1 - x_n)^2}}$$

and

$$I_2 = \frac{V}{Z_2} = \frac{V}{2R} \frac{1 + x_n}{\sqrt{[Q - (Q_0 - Q)x_n]^2 + (1 + x_n)^2}}$$

in place of instantaneous currents i_1 and i_2 in the first form of Equation (2-6). The flux densities in the gaps are

$$\mathscr{B}_1 = \frac{\mu N I_1}{g_0(1 - x_n)}$$

and

$$\mathscr{B}_2 = \frac{\mu N I_2}{g_0(1 + x_n)}.$$

The currents can be normalized with respect to the resonant current

$$I_R = \frac{V}{2R},$$

so that

$$I_{n1} = \frac{I_1}{I_R} = \frac{1 - x_n}{\sqrt{[Q + (Q_0 - Q)x_n]^2 + (1 - x_n)^2}} = \frac{1}{\sqrt{1 + \left(Q + \frac{Q_0 x_n}{1 - x_n}\right)^2}}$$

and

$$I_{n2} = \frac{I_2}{I_R} = \frac{1 + x_n}{\sqrt{[Q - (Q_0 - Q)x_n]^2 + (1 + x_n)^2}} = \frac{1}{\sqrt{1 + \left(Q - \frac{Q_0 x_n}{1 + x_n}\right)^2}}$$

and the flux densities can be expressed in terms of these normalized currents:

$$\mathscr{B}_1 = \frac{\mu N I_{n1} I_R}{g_0(1 - x_n)}$$

and

$$\mathscr{B}_2 = \frac{\mu N I_{n2} I_R}{g_0(1 + x_n)}$$

or in normalized form,

$$\mathscr{B}_{n1} = \frac{I_{n1}}{1 - x_n}$$

and

$$\mathscr{B}_{n2} = \frac{I_{n2}}{1 + x_n}.$$

Thus the normalized currents and flux densities are simply related through the normalized displacements.

2-4 Conditions for Stable Operation

For the block to be in stable equilibrium in its central position, the numerator of Equation (2-8) must be negative; in other words, the magnetic force that accompanies displacement must be in the opposite direction from the direction of displacement. Hence the criterion for stable operation is

$$Q_0 Q > Q^2 + 1 \qquad \qquad (2\text{-}9)$$

or

$$Q_0 > Q + \frac{1}{Q}.$$

As can be readily shown, the sum of a number (here represented by Q) and its reciprocal is a minimum when $Q = 1$, so that the lower limit of Q_0 for stable operation is

$$Q_0 > 2. \qquad \qquad (2\text{-}10)$$

As Q becomes very small or very large, Q_0 must become very large to have stable operation. Conversely, for any fixed $Q_0 > 2$, stable operation requires that

$$\tfrac{1}{2}(Q_0 - \sqrt{Q_0^2 - 4}) < Q < \tfrac{1}{2}(Q_0 + \sqrt{Q_0^2 - 4}). \qquad \qquad (2\text{-}11)$$

2-5 Quiescent Currents

Examination of Figure 2-2 indicates that the behavior of the system when the block is displaced is considerably determined by the settings of the currents with respect to the resonance curve when the block is centered. This setting is in fact determined by the quality factor Q as fixed for any one device by the amount of tuning capacitance C. The current setting commonly is expressed in terms of the *power point*, or the ratio of the power dissipated by the quiescent current to the power dissipated by the resonant current. The quiescent current in one circuit is

$$I_0 = \frac{V}{2R\sqrt{Q^2 + 1}} = \frac{I_R}{\sqrt{Q^2 + 1}} \tag{2-12}$$

so that the power point is

$$\frac{I_0^2}{I_R^2} = \frac{1}{Q^2 + 1}. \tag{2-13}$$

For the circuit under discussion, operation from the half-power point generally is found to be practically advantageous, for reasons mentioned presently. Then $Q = 1$, and the equation for normalized force becomes

$$F_n = -\frac{4(Q_0 - 2)x_n}{4 + 8(Q_0 - 1)x_n^2 + [1 + (Q_0 - 1)^2]^2 x_n^4}. \tag{2-14}$$

Unity power point corresponds to resonance, for which the system is unstable. The correspondence between power point and Q is tabulated in Table 2-1.

Table 2-1 Correspondence between Power Point and Q

Power Point	Q	Power Point	Q
1.0	0.	0.4	1.23
0.9	0.333	$0.333 = 1/3$	$1.41 = \sqrt{2}$
0.8	0.500	0.3	1.53
$0.75 = 3/4$	$0.578 = 1/\sqrt{3}$	$0.25 = 1/4$	$1.73 = \sqrt{3}$
0.7	0.655	0.2	2.00
$0.667 = 2/3$	$0.707 = 1/\sqrt{2}$	0.1	3.00
0.6	0.817	$0.0588 = 1/17$	4.00
0.5	1.000	0.	∞

In Figures 2-3 through 2-11 are shown families of curves of F_n, I_n, and $\mathscr{B}_n$ versus x_n for various combinations of Q and Q_0 taken from digital computer solutions. Here $Q_0 = 10$ is fairly representative of practical operation, and $Q_0 = 6$ and $Q_0 = 16$ are the practical extremes. In general, $Q = 1$ is representative of practical operation, and $Q = \sqrt{3}$ and $Q = 1/\sqrt{3}$ might be the practical extremes, a range from $\frac{1}{4}$ to $\frac{3}{4}$ power point.

2-6 Maximum or Peak Force and Stiffness

Of primary interest in the application of passive suspensions are the stiffness, or slope of the force-displacement curve at the origin, and the force peak and the displacement at which it occurs. The stiffness is of interest for erasing small displacements of the suspended member, and the force peak and its location are of interest for erecting the member from its bearings or rest stops when the suspension is energized. The derivative of the force-displacement curve with respect to displacement is too complicated to permit solution for zero slope and thus locate the peak. The peaks of the force curves are most readily obtained by examination of Figures 2-3 through 2-5. The stiffness at zero displacement, sometimes called the *initial stiffness*, can be found very simply by dividing the force-displacement equation by x and then letting x approach zero:

$$\frac{dF_{Vs}}{dx}\bigg|_{x\to 0} = \dot{F}_{Vs}\bigg|_{x\to 0} = \frac{F_{Vs}}{g_0 x_n}\bigg|_{x_n\to 0} = \frac{N^2 V^2 \mu A}{4R^2 g_0^3} \frac{4(Q^2 - Q_0 Q + 1)}{(Q^2 + 1)^2} \tag{2-15}$$

or, in normalized form,

$$\dot{F}_{n0} = \frac{\dot{F}_{Vs}}{\dot{F}_V}\bigg|_{x\to 0} = \frac{4(Q^2 - Q_0 Q + 1)}{(Q^2 + 1)^2}. \tag{2-16}$$

This function is plotted in Figure 2-12, and in Figure 2-13 the practically useful part is considerably expanded. The stability criteria derived in Section 2-4 of course apply to stiffness also, as is apparent from this equation. Only those parts of Figure 2-12 that are negative and correspond to positive Q represent stable operating conditions. Differentiating Equation (2-16) with respect to Q and equating to zero gives

$$\frac{d\dot{F}_{n0}}{dQ} = \ddot{F}_{n0} = \frac{-2Q^3 + 3Q_0 Q^2 - 2Q - Q_0}{(Q^2 + 1)^3} = 0,$$

so that for a mathematical maximum or minimum of stiffness at $x = 0$,

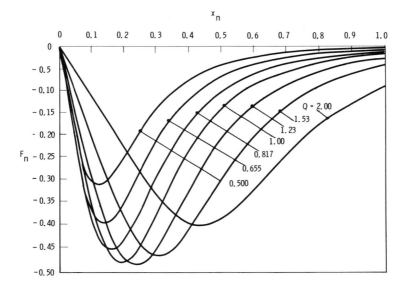

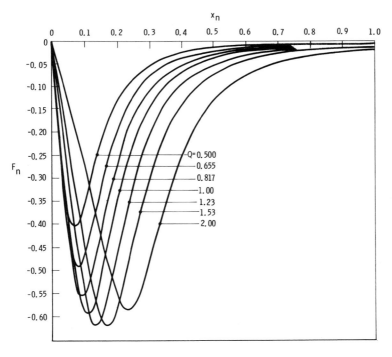

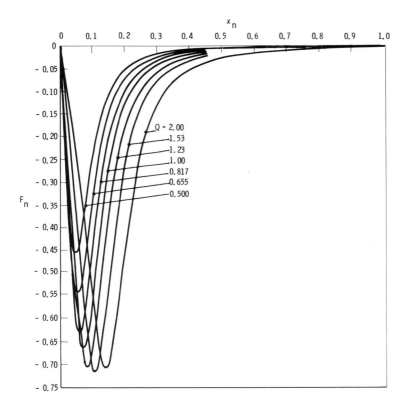

Figure 2-3 Normalized centering force versus normalized displacement, Case (1), or Case (7) (Chapter 4); $Q_0 = 6$.

Figure 2-4 Normalized centering force versus normalized displacement, Case (1), or Case (7) (Chapter 4); $Q_0 = 10$.

Figure 2-5 Normalized centering force versus normalized displacement, Case (1), or Case (7) (Chapter 4); $Q_0 = 16$.

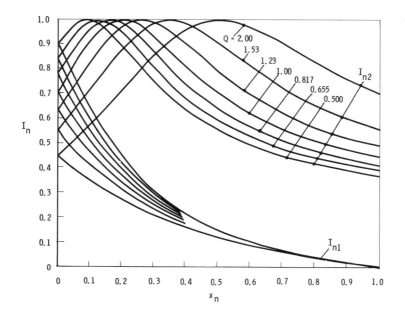

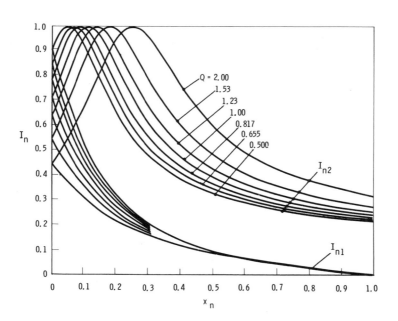

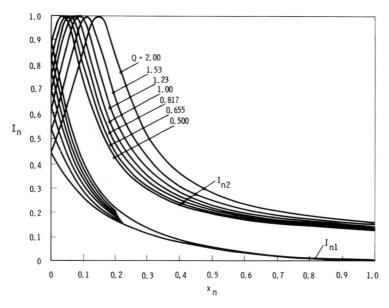

Figure 2-6 Normalized coil currents versus normalized displacement, Case (1), or Case (7) (Chapter 4); $Q_0 = 6$.

Figure 2-7 Normalized coil currents versus normalized displacement, Case (1), or Case (7) (Chapter 4); $Q_0 = 10$.

Figure 2-8 Normalized coil currents versus normalized displacement, Case (1), or Case (7) (Chapter 4); $Q_0 = 16$.

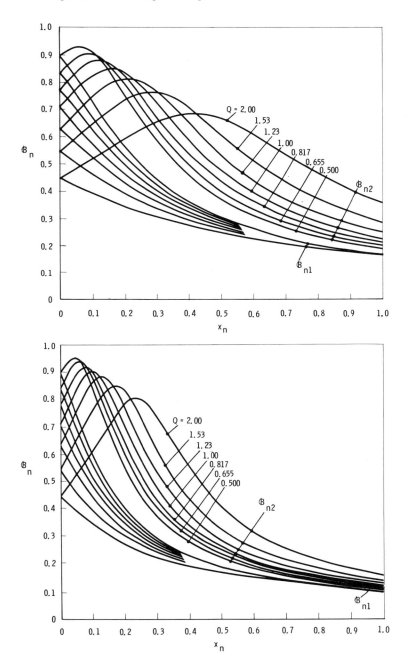

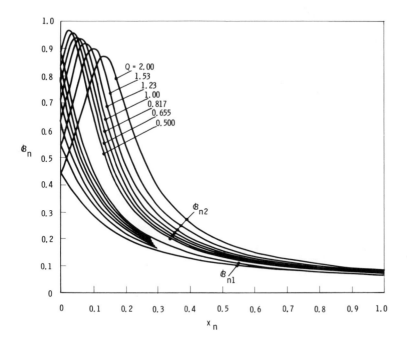

Figure 2-9 Normalized flux densities versus normalized displacement, Case (1), or Case (7) (Chapter 4); $Q_0 = 6$.

Figure 2-10 Normalized flux densities versus normalized displacement, Case (1), or Case (7) (Chapter 4); $Q_0 = 10$.

Figure 2-11 Normalized flux densities versus normalized displacement, Case (1), or Case (7) (Chapter 4); $Q_0 = 16$.

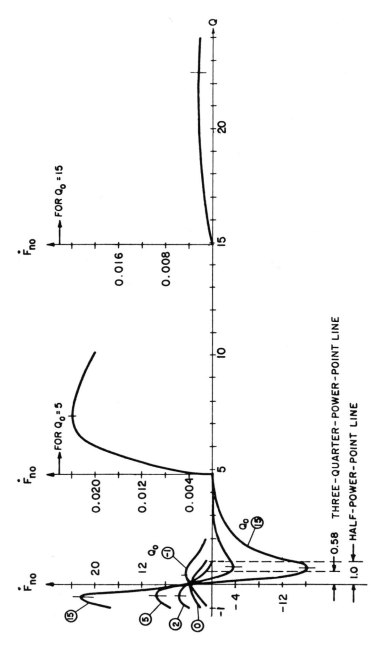

Figure 2-12 Normalized initial stiffness versus Q for various values of Q_0, Case (1), or Case (7) (Chapter 4).

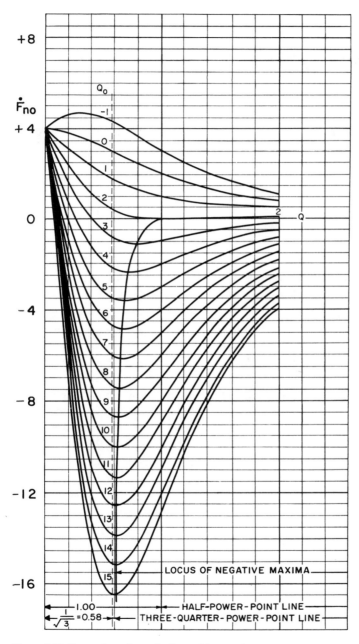

Figure 2-13 Expansion of practically useful part of Figure 2-12.

$$Q_0 = \frac{2Q(Q^2 + 1)}{3Q^2 - 1},$$
(2-17)

and the corresponding peak normalized stiffnesses are

$$\dot{F}_{nm} = \frac{4(Q^2 - 1)}{(3Q^2 - 1)(Q^2 + 1)}.$$
(2-18)

Three values of Q and hence three values of F_{nm} correspond to each Q_0, as marked on Figure 2-12 for the $Q_0 = 5$ and $Q_0 = 15$ curves. For $Q = 1$, *peak* stiffness $\dot{F}_{nm}$ is zero and occurs on the $Q_0 = 2$ curve; for $Q > 1$, the peak stiffness $\dot{F}_{nm}$ is positive and quite small, as illustrated for the $Q = 5$ and $Q = 15$ curves; for $0 < Q < 1/\sqrt{3}$, the peak stiffness $\dot{F}_{nm}$ is positive and occurs on curves for $Q_0 < 2$; for $Q < 0$, the peak stiffness is positive. Hence only in the region for which

$$1/\sqrt{3} < Q < 1$$
(2-19)

is peak stiffness negative. Equations (2-17) and (2-18) are parametric equations that define the loci of peak stiffnesses $\dot{F}_{nm}$ in the $\dot{F}_{n0}$-Q plane. In Figure 2-13 the locus is shown for the region of Equation (2-19); the locus starts at $\dot{F}_{nm} = 0$, $Q = 1$, on line $Q_0 = 2$ and increases in magnitude as Q_0 increases, approaching $\dot{F}_{nm} = -\infty$, $Q = 1/\sqrt{3}$, on line $Q_0 = \infty$. Given any Q_0, the intersection of the Q_0 curve with the locus gives the corresponding $\dot{F}_{nm}$ and Q; given any Q, the corresponding $\dot{F}_{nm}$ and Q_0 can be found from the locus; given any $\dot{F}_{nm}$, the corresponding Q_0 and Q can be found from the locus.

For a fixed $Q_0 > 3.34$, the stiffness $\dot{F}_{n0}$ for three-quarter-power-point operation always is somewhat larger than for half-power-point operation. Practical suspensions commonly have Q_0 in the range 5 to 20. Hence, from the standpoint of stiffness at the origin, the objective should be to operate close to the three-quarter power point and at as high Q_0 as possible. However, as examination of the force plots of Figures 2-3 through 2-5 shows, for given Q_0, the peak force is smaller for three-quarter-power-point operation than for half-power-point operation and occurs at smaller displacement. Further, and perhaps more important, the quiescent voltage across the coils is

$$V_{RL} = V \sqrt{\frac{Q_L^2 + 1}{Q^2 + 1}},$$
(2-20)

in which

$$Q_L = \frac{\omega(L_\ell + L_0)}{R} = \frac{\omega L_\ell}{R} + Q_0 = Q + Q_c.$$

With L_ℓ sometimes as high as $0.5L_0$, Q_L^2 can be more than double Q_0^2, so that high Q_0 and low Q can make the voltage V_{RL} across the coil many times the source voltage V, and with the rather sharp tuning that corresponds to high Q_L and low Q the voltage across the capacitor would be essentially the same as the voltage across the coil. As the block departs from the central position, the voltages across the coil and the capacitor at the end having the longer gap increase. Hence owing to the development of excessive circuit voltages, use of extremely high Q_0 would not be practical even if achievable, and operation in the vicinity of the half-power point is preferable, partly to avoid excessive circuit voltages and particularly to avoid a decrease in peak force and a decrease in the displacement at which it occurs.

2-7 Position Signals

The two operating circuits of Figure 2-1 can be used as an impedance bridge circuit, the unbalance of which, indicated by voltage V_B, can be taken as a measure of the displacement of the block. In fact, such a bridge circuit can be used as a means of providing the error signal for servo control of an active suspension. Use of such a bridge circuit in association with the early active suspensions, as described in Chapter 9, was a circumstance that led to the development of the passive type of suspension under consideration. Solution of the bridge circuit of Figure 2-1 gives for the normalized position signal

$$V_n = \frac{V_B}{V} = \frac{j(1 + x_n)(Q_L - Q)}{(1 + x_n) + j[Q - (Q_0 - Q)x_n]}$$
$$- \frac{j(1 - x_n)(Q_L - Q)}{(1 - x_n) + j[Q + (Q_0 - Q)x_n]}, \tag{2-21}$$

which for very small displacements reduces to

$$|V_n| \approx \frac{2Q_0(Q_L - Q)x_n}{Q^2 + 1}, \tag{2-22}$$

and for half-power-point operation reduces to

$$|V_n| \approx Q_0(Q_L - 1)x_n. \tag{2-23}$$

These signals can be used to indicate that the block actually is suspended and to adjust its position.

2-8 Initial Positioning by Trimming Capacitance

Owing to slight differences in resistances, possible slight differences in numbers of coil turns, small geometric dissymmetries, and magnetic material that is not exactly homogeneous and isotropic, conditions that are practically inevitable to some extent in physical apparatus, the block may not be exactly centered when conditions are quiescent. The position of the *geometric center* of the suspended member with respect to the *geometric center* of the stationary or reference member in the quiescent state sometimes is called the *force center*, but it does not necessarily indicate a point through which a resultant force acts. The block may be centered by trimming the capacitances.

Alternatively, a slight off-center quiescent condition may be desirable, for example, to center a gyro float with respect to its bearings (with a two- or three-axis suspension) rather than center the rotor of the suspension with respect to the stator. This condition can be established also by trimming the capacitances, whether appreciable imperfections exist in the structure or not. The parameter relations for this trimming operation can be examined fairly simply for the single-axis suspension, and apply nearly enough for small irregularities or displacements of two- and three-axis suspensions.

A most likely situation is unequal coil resistances. If these resistances are

$$R_1 = R - r$$

and

$$R_2 = R + r$$

in Figure 2-1 and the corresponding capacitances are

$$C_1 = C - c$$

and

$$C_2 = C + c$$

the corresponding gaps may be taken as

$$g_{10} = g_0 - x_0$$

and

$$g_{20} = g_0 + x_0.$$

Here r represents the resistance deviation from the mean coil resistance R, c represents the trimming capacitance with respect to the tuning capacitance

C for a perfect structure in its central position, and x_0 represents the deviation from the geometric center.

For force balance with offset x_0, the flux densities (with negligible fringing) must be the same at all poles, which means that

$$\left|\frac{I_1}{I_2}\right| = \frac{g_0 - x_0}{g_0 + x_0} = \left|\frac{Z_2}{Z_1}\right|$$

$$= \sqrt{\frac{[2R(1 + r_n)]^2 + \left[2\omega\left(L - \frac{L_0 x_{n0}}{1 + x_{n0}}\right) - \frac{1}{\omega C(1 + c_n)}\right]^2}{[2R(1 - r_n)]^2 + \left[2\omega\left(L + \frac{L_0 x_{n0}}{1 - x_{n0}}\right) - \frac{1}{\omega C(1 - c_n)}\right]^2}} \tag{2-24}$$

or

$$\frac{(1 - x_{n0})^2}{(1 + x_{n0})^2} = \frac{(1 + r_n)^2 + \left|Q_L - \frac{Q_0 x_{n0}}{1 + x_{n0}} - \frac{Q_L - Q}{1 + c_n}\right|^2}{(1 - r_n)^2 + \left|Q_L + \frac{Q_0 x_{n0}}{1 - x_{n0}} - \frac{Q_L - Q}{1 - c_n}\right|^2}, \tag{2-25}$$

in which Q_L and Q_0 are based on R, Q is based on R and C, and

$$x_{n0} = \frac{x_0}{g_0},$$

$$r_n = \frac{r}{R},$$

and

$$c_n = \frac{c}{C}.$$

Solution of Equation (2-25) for small displacements gives

$$x_{n0} \approx \frac{r_n + (Q_L - Q)Q c_n}{(Q_0 - Q)Q - 1} \tag{2-26}$$

or

$$c_n \approx \frac{[(Q_0 - Q)Q - 1]x_{n0} - r_n}{(Q_L - Q)Q}. \tag{2-27}$$

These equations permit estimates of trimming capacitance c required to give a desired offset x_0, to center the block ($x_0 = 0$), or to compute x_0 from known r and c. For the special case of capacitance C originally set for half-power-point operation with resistance R, the equations become, respectively,

$$x_{n0} \approx \frac{r_n + (Q_L - 1)c_n}{Q_0 - 2} \tag{2-28}$$

and

$$c_n \approx \frac{x_{n0}(Q_0 - 2) - r_n}{Q_L - 1}. \tag{2-29}$$

The relations between offset x_0 and trimming capacitance c for a perfect suspension can be obtained by letting $r = 0$ in Equations (2-26) through (2-29).

With these small adjustments for quiescent conditions, the force-displacement relations for the two ends are slightly different, but the general behavior of the device is not changed appreciably. The actual force-displacement equation can be derived, if desirable, by separate derivation of the force relations for the two ends from Equation (2-6). The application of this capacitive trimming to two- and three-axis suspensions is considered in subsequent chapters. For example, the alignment of the axis of a gyro float that has a suspension at each end can be controlled by trimming the capacitances associated with the suspensions at the two ends.

When the block is located at the force center, the signal output of the bridge circuit is not necessarily zero; that is, the *signal center*, or position for zero signal, and the force center do not necessarily coincide, and the trimming process for adjustment of the force center does not necessarily carry with it a corresponding adjustment of the signal center. The distance between the force and signal centers is called the *suspension position error*. The same imperfections influence both the departure of the force center and the signal center from the geometric center but not in the same way, so that circuit adjustments do not bring coincidence. Signal voltage is a function of flux, whereas force is a function of flux squared. The slight inequities in resistance and capacitance give a position signal for the quiescent position with displacement x_0 that is a function of r and c. If this signal is bucked out, the additional position signal from the bridge circuit of Figure 2-1 is proportional to the displacement x measured from x_0 exactly as given in Equation (2-22) for very small displacements.

Once a trimming adjustment is made, it does not necessarily stay put but is somewhat unstable, as influenced by small random drifts in parameter values in the tuning or control circuitry. This random wandering of the force and signal centers over a period of time must stay within specified limits for a particular application to ensure satisfactory operation.

In Equation (2-24) no allowance is made for possible different cross-sectional areas or gap lengths of a pole pair, which would greatly complicate the algebra, nor can irregularities in the properties of the magnetic materials be taken into account. The resulting equations, (2-26) through (2-29), should give useful first estimates for the trimming procedure, which then must be refined by trial.

2-9 Summary
The series tuned-circuit arrangement designated as Case (1) probably has been the most commonly used scheme for passive suspension. By generalizing the relations that characterize performance, the considerable range of tuning that theoretically is possible within stable operation is fully displayed. Plots of coil currents, gap flux densities, and restoring forces as functions of displacement are illustrative of the performance. However, the actual stiffness and force characteristics practically achievable are limited by magnetic saturation and the circuit currents and voltages that can be tolerated. These limitations are discussed in subsequent chapters. The position of the suspended member can be determined by signals obtained from the circuitry itself, and the quiescent position can be adjusted by circuit parameter trimming, usually capacitive.

3 OTHER TUNED-CIRCUIT CONNECTIONS FOR SINGLE-AXIS MAGNETIC SUSPENSIONS

Use of a current source instead of a voltage source with series tuning and use of either a current source or a voltage source with parallel tuning give magnetic suspensions of somewhat different characteristics, even though the magnetic structure and the windings are unchanged. These differences are illustrated by further plots of forces, currents, flux densities, and stiffnesses but, for economy of space and computer time, with fewer plots for each circuit arrangement than given in Chapter 2 for Case (1). Additional plots can be derived from the plots given by means of interrelationships derived.

3-1 Current and Voltage Sources

In Chapter 2 the force and stiffness relations of the single-axis magnetic suspension are derived on the basis of the circuitry of Figure 2-1, which shows an ideal voltage source. A physical source contains internal impedance, so that the actual behavior of the suspension is modified to the extent that the total current taken by the suspension affects the terminal voltage of the source. As shown by Equations (2-6), (2-7), and (2-15), the restoring force and stiffness of the suspension depend on the square of that voltage. Since in operation the current of one circuit increases and the current of the other circuit decreases when the block is displaced, as shown in Figure 3-1, the change in terminal voltage is not large if the internal series impedance of the source is small compared with the impedance presented by the suspension circuits. In experimental testing, the terminal voltage must be kept adjusted to a fixed value to simulate voltage-source operation.

If the internal series impedance of the source is large compared with the impedance presented by the suspension circuits, current-source operation is approximated and the force and stiffness relations change. In experimental

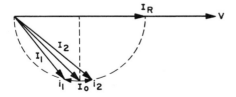

Figure 3-1 Changes in phasor currents, Case (1), with block off center.

testing the source current must be kept adjusted to the same value to simulate constant-current operation.

3-2 Current Source, Series Tuning, Case (2)

For this type of operation, Figure 2-1 can be used with replacement of voltage source V by current source I. Use of Equation (2-6) leads to

$$F_{Is} = N^2\mu A\left[\frac{I_1^2}{(g_0 - x)^2} - \frac{I_2^2}{(g_0 + x)^2}\right]. \tag{3-1}$$

Substitution of

$$I_1 = \frac{Z_2}{Z_1 + Z_2}I = \frac{I(1 - x_n)}{2}\sqrt{\frac{[Q - (Q_0 - Q)x_n]^2 + (1 + x_n)^2}{[Q + (Q_0 - Q)x_n^2]^2 + (1 - x_n^2)^2}}$$

$$= \frac{I}{2}\sqrt{\frac{1 + \left(Q - \dfrac{Q_0 x_n}{1 + x_n}\right)^2}{1 + \left(Q + \dfrac{Q_0 x_n^2}{1 - x_n^2}\right)^2}}$$

and

$$I_2 = \frac{Z_1}{Z_1 + Z_2}I = \frac{I(1 + x_n)}{2}\sqrt{\frac{[Q + (Q_0 - Q)x_n]^2 + (1 - x_n)^2}{[Q + (Q_0 - Q)x_n^2]^2 + (1 - x_n^2)^2}}$$

$$= \frac{I}{2}\sqrt{\frac{1 + \left(Q + \dfrac{Q_0 x_n}{1 - x_n}\right)^2}{1 + \left(Q + \dfrac{Q_0 x_n^2}{1 - x_n^2}\right)^2}}$$

leads to

$$F_{Is} = \frac{N^2 I^2 \mu A(Q^2 - Q_0 Q + 1)x_n}{g_0^2\{[Q + (Q_0 - Q)x_n^2]^2 + (1 - x_n^2)^2\}} \tag{3-2}$$

or

$$F_n = \frac{F_{Is}}{F_I} = \frac{4(Q^2 - Q_0 Q + 1)x_n}{[Q + (Q_0 - Q)x_n^2]^2 + (1 - x_n^2)^2}, \tag{3-3}$$

in which the base for normalizing,

$$F_I = \frac{N^2 I^2 \mu A}{4g_0^2},$$

is the pull on one side of the block when it is centered. The normalized force expressions for operation with current source and with voltage source, Cases (2) and (1), are related by

$$\frac{F_{nI}}{F_{nV}} = \frac{(Z_1 Z_2)^2}{R^2(Z_1 + Z_2)^2} = \frac{1}{R^2(Y_1 + Y_2)^2} \tag{3-4}$$

if $V = RI$, and the actual average forces are related by

$$\frac{F_{Is}}{F_{Vs}} = \frac{I^2}{V^2} \frac{(Z_1 Z_2)^2}{(Z_1 + Z_2)^2}. \tag{3-5}$$

For operation with a current source, Case (2), Q has not the same significance in terms of power point as for operation from a voltage source, Case (1), because the quiescent power is fixed when a current source is used, but Q can serve as a parameter indicative of the setting of capacitance C. For $Q = 1$,

$$F_n = \frac{-4(Q_0 - 2)x_n}{[1 + (Q_0 - 1)x_n^2]^2 + (1 - x_n^2)^2}. \tag{3-6}$$

If normalized currents are taken as

$$I_{n1} = \frac{2I_1}{I}$$

and

$$I_{n2} = \frac{2I_2}{I},$$

gap flux densities are

$$\mathcal{B}_1 = \frac{\mu N I_{n1} I}{2g_0(1 - x_n)}$$

and

$$\mathcal{B}_2 = \frac{\mu N I_{n2} I}{2g_0(1 + x_n)},$$

or, in normalized form,

$$\mathscr{B}_{n1} = \frac{I_{n1}}{1 - x_n}$$

and

$$\mathscr{B}_{n2} = \frac{I_{n2}}{1 + x_n},$$

as for Case (1), with due regard to the bases for normalization. Normalized curves of F_n, I_n, and $\mathscr{B}_n$ versus x_n, are plotted in Figures 3-2 through 3-8 for various combinations of Q_0 and Q. These plots should be compared with Figures 2-3 through 2-11. The static stability conditions are the same as for operation from a voltage source, Section 2-4.

With the block centered, the stiffness of the suspension is

$$\dot{F}_{Is}\bigg|_{x \to 0} = \frac{N^2 I^2 \mu A}{4 g_0^3} \frac{4(Q^2 - Q_0 Q + 1)}{Q^2 + 1} \tag{3-7}$$

or, in normalized form,

$$\dot{F}_{n0} = \frac{4(Q^2 - Q_0 Q + 1)}{Q^2 + 1} = -4\left|\frac{Q_0}{Q + \dfrac{1}{Q}} - 1\right|. \tag{3-8}$$

The stiffnesses for operation with current source and with voltage source for series tuning, Cases (2) and (1), are related by

$$\frac{\dot{F}_{Is}}{\dot{F}_{Vs}}\bigg|_{x \to 0} = \frac{R^2 I^2}{V^2}(Q^2 + 1). \tag{3-9}$$

Hence if $V = RI$, the stiffness when the block is centered is $(Q^2 + 1)$ times as much for operation with a current source, Case (2), as for operation with a voltage source, Case (1). Differentiation of Equation (3-7) and equating to zero shows that for any value of Q_0 maximum initial stiffness occurs with $Q = 1$. In normalized form, it is

$$\dot{F}_{nm} = -2(Q_0 - 2). \tag{3-10}$$

In Figure 3-9, normalized stiffness $\dot{F}_{n0}$ is plotted against Q for various values of Q_0. This plot should be compared with Figure 2-13.

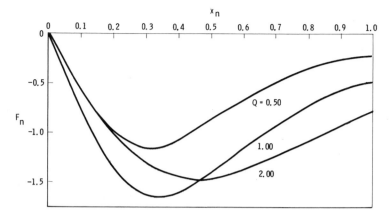

Figure 3-2 Normalized centering force versus normalized displacement, Case (2), or Case (6) (Chapter 4); $Q_0 = 6$.

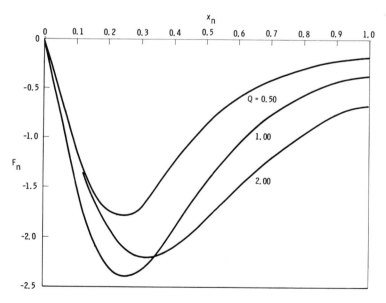

Figure 3-3 Normalized centering force versus normalized displacement, Case (2), or Case (6) (Chapter 4); $Q_0 = 10$.

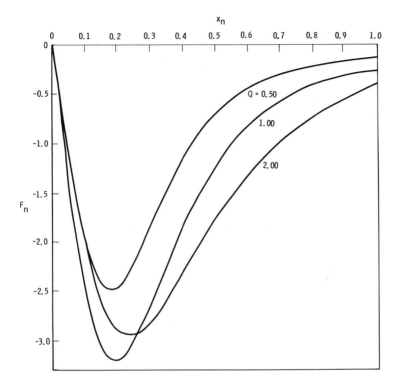

Figure 3-4 Normalized centering force versus normalized displacement, Case (2), or Case (6) (Chapter 4); $Q_0 = 16$.

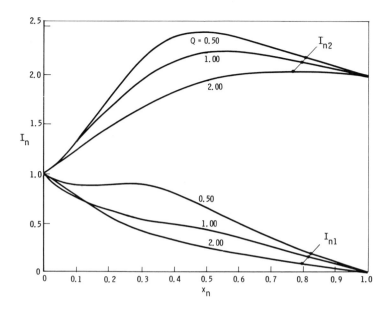

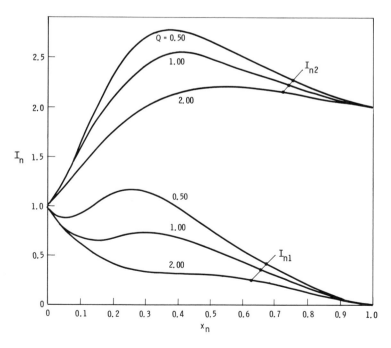

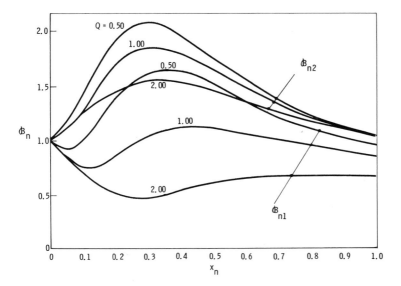

Figure 3-5 Normalized coil currents versus normalized displacement, Case (2), or Case (6) (Chapter 4); $Q_0 = 6$.

Figure 3-6 Normalized coil currents versus normalized displacement, Case (2), or Case (6) (Chapter 4); $Q_0 = 10$.

Figure 3-7 Normalized flux densities versus normalized displacement, Case (2), or Case (6) (Chapter 4); $Q_0 = 10$.

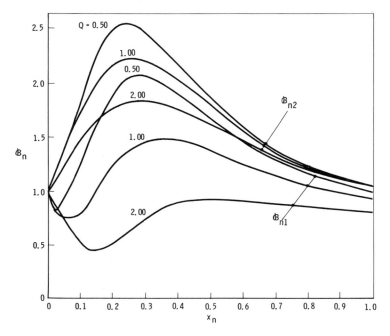

Figure 3-8 Normalized flux densities versus normalized displacement, Case (2), or Case (6) (Chapter 4); $Q_0 = 16$.

The quiescent voltages across the coils and capacitor are

$$V_{RL} = RI\sqrt{Q_L^2 + 1} \tag{3-11}$$

and

$$V_C = \frac{I}{2\omega C} = RI(Q_L - Q), \tag{3-12}$$

which for sharp tuning corresponding to high Q_L and low Q can be quite high and can increase at the end having the longer gap, as the block is displaced and resonance is approached.

Position signals can be obtained from the bridge-circuit output of Figure 2-1 with I substituted for V:

$$V_B = \frac{I_2 - I_1}{j\omega C} = \frac{(Z_1 - Z_2)I}{j\omega C(Z_1 + Z_2)} = \frac{2RIQ_0(Q_L - Q)x_n}{1 - x_n^2 + j[Q + (Q_0 - Q)x_n^2]} \tag{3-13}$$

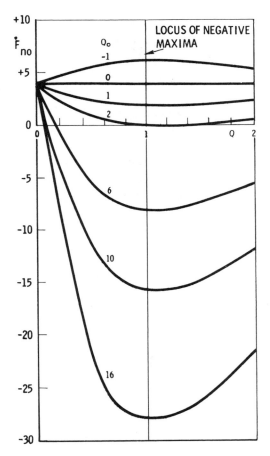

Figure 3-9 Normalized initial stiffness versus Q, for various values of Q_0, Case (2), or Case (6) (Chapter 4).

or

$$|V_n| = \left|\frac{V_B}{RI}\right| = \frac{2Q_0(Q_L - Q)x_n}{\sqrt{[Q + (Q_0 - Q)x_n^2]^2 + (1 - x_n^2)^2}} \approx \frac{2Q_0(Q_L - Q)x_n}{\sqrt{Q^2 + 1}}, \quad (3\text{-}14)$$

the final form being for very small displacement.

The conditions for positioning by means of trimming capacitance are the same as shown in Section 2-8.

3-3 Voltage or Current Source, Parallel Tuning,[32] Cases (3) and (4)

Owing to certain algebraic complexities, these two cases are advantageously considered together. For these connections, a tuning capacitor is placed in parallel with the operating coils at each end of the block, as shown in Figure 3-10 (with G omitted), and the two coil-capacitor combinations are placed in series. In terms of rms magnitudes, the average force component of Equation (2-6) can be written as

$$F_p = N^2\mu A\left(\frac{I_{L1}^2}{g_1^2} - \frac{I_{L2}^2}{g_2^2}\right) = N^2\mu A\left(\frac{V_1^2}{g_1^2 Z_{L1}^2} - \frac{V_2^2}{g_2^2 Z_{L2}^2}\right), \quad (3\text{-}15)$$

in which I_{L1} and I_{L2} are the respective coil currents,

$$Z_{L1} = 2[R + j\omega(L_\ell + L_1)],$$

and

$$Z_{L2} = 2[R + j\omega(L_\ell + L_2)].$$

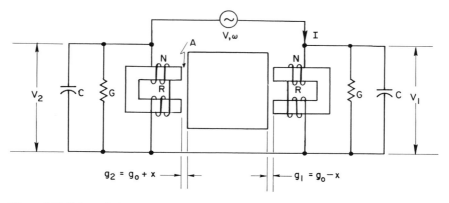

Figure 3-10 Schematic for single-axis magnetic suspension, Cases (3), (4), (3a), and (4a).

Since

$$\frac{V_1^2}{V^2} = \frac{Z_1^2}{(Z_1 + Z_2)^2} = \frac{Y_2^2}{(Y_1 + Y_2)^2}$$

$$\frac{V_2^2}{V^2} = \frac{Z_2^2}{(Z_1 + Z_2)^2} = \frac{Y_1^2}{(Y_1 + Y_2)^2},$$

$$\frac{I_{L1}^2}{I^2} = \frac{Y_{L1}^2}{Y_1^2} = \frac{Z_1^2}{Z_{L1}^2},$$

and

$$\frac{I_{L2}^2}{I^2} = \frac{Y_{L2}^2}{Y_2^2} = \frac{Z_2^2}{Z_{L2}^2},$$

Equation (3-15) can be written as

$$\begin{aligned}
F_{Vp} &= N^2 V^2 \mu A \left[\frac{Z_1^2}{g_1^2 Z_{L1}^2 (Z_1 + Z_2)^2} - \frac{Z_2^2}{g_2^2 Z_{L2}^2 (Z_1 + Z_2)^2} \right] \\
&= N^2 V^2 \mu A \left[\frac{1}{g_1^2 Z_{L1}^2 (1 + Y_1 Z_2)^2} - \frac{1}{g_2^2 Z_{L2}^2 (1 + Y_2 Z_1)^2} \right]
\end{aligned} \tag{3-16}$$

for Case (3) or as

$$F_{Ip} = N^2 I^2 \mu A \left[\frac{Y_{L1}^2}{g_1^2 Y_1^2} - \frac{Y_{L2}^2}{g_2^2 Y_2^2} \right] = N^2 I^2 \mu A \left[\frac{1}{g_1^2 Y_1^2 Z_{L1}^2} - \frac{1}{g_2^2 Y_2^2 Z_{L2}^2} \right] \tag{3-17}$$

for Case (4), in which

$$Y_1 = j\omega C + Y_{L1} = j\omega C + \frac{1}{Z_{L1}}$$

and

$$Y_2 = j\omega C + Y_{L2} = j\omega C + \frac{1}{Z_{L2}}.$$

Substitution of the expressions for the various admittances and impedances into the second form of Equation (3-16) gives, for Case (3),

$$F_{Vp} = \frac{N^2 V^2 \mu A}{4R^2 g_0^2} \frac{(Q^2 - Q_0 Q + 1) x_n}{\{Q_L + Q - [(Q_L - Q_0) - (Q_0 - Q)] x_n^2\}^2 + \{Q_L Q - 1 + [(Q_L - Q_0)(Q_0 - Q) + 1] x_n^2\}^2} = \frac{N^2 V^2 \mu A}{4R^2 g_0^2} F_n. \tag{3-18}$$

This form of the equation has the same normalizing factor as used for series tuning, Case (1), and the factor of 4 missing in the numerator of the normalized part of Equation (3-18) could be retrieved by using V for half the applied voltage, or the quiescent voltage across a coil pair, but such a move seems to be without much merit. Use of this normalizing factor in Equation (3-18) permits direct comparison with the force for series tuning with voltage source, Case (1), though here the factor does not in itself represent any particular force. The normalized coil currents, on the same base as for Case (1), can be obtained from

$$I_{L1} = \frac{V}{2R}\frac{(1-x_n)}{2}\sqrt{\frac{[Q-(Q_0-Q)x_n]^2+(1+x_n)^2}{\{Q_L+Q-[(Q_L-Q_0)-(Q_0-Q)]x_n^2\}^2 + \{Q_LQ-1+[(Q_L-Q_0)(Q_0-Q)+1]x_n^2\}^2}} = \frac{V}{2R}I_{nL1}$$

and

$$I_{L2} = \frac{V}{2R}\frac{(1+x_n)}{2}\sqrt{\frac{[Q+(Q_0-Q)x_n]^2+(1-x_n)^2}{\{Q_L+Q-[(Q_L-Q_0)-(Q_0-Q)]x_n^2\}^2 + \{Q_LQ-1+[(Q_L-Q_0)(Q_0-Q)+1]x_n^2\}^2}} = \frac{V}{2R}I_{nL2},$$

and the relation between normalized currents and normalized flux densities is as derived for the previous cases. Normalized curves for F_n, I_{nL}, and $\mathcal{B}_n$ versus x_n are plotted in Figures 3-11 through 3-13, for $Q_L = 15$ and $Q_0 = 10$, F_n being plotted first to the same scale as for Case (1), then to a readable scale. Curves for other practical Q_L, Q_0 combinations are not shown, because the differences are minor, and because, as discussed in Chapter 5, this connection is not especially practical for constant-voltage operation. These plots should be compared with corresponding plots for other cases.

The stiffness at zero displacement is

$$\dot{F}_{Vp}\Big|_{x\to 0} = \frac{N^2V^2\mu A}{4R^2g_0^3}\frac{Q^2 - Q_0Q + 1}{(Q_L^2 + 1)(Q^2 + 1)} = \frac{N^2V^2\mu A}{4R^2g_0^3}\dot{F}_{n0}. \qquad (3\text{-}19)$$

Hence the stiffnesses *for equal voltage sources*, for parallel and series tuning, Cases (3) and (1), are related by

$$\frac{\dot{F}_{Vp}}{\dot{F}_{Vs}}\Big|_{x\to 0} = \frac{Q^2 + 1}{4(Q_L^2 + 1)}. \qquad (3\text{-}20)$$

Evidently, the series arrangement, Case (1), Figure 2-1, is much the stiffer

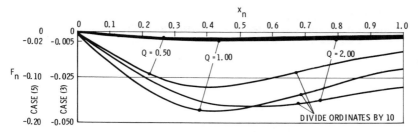

Figure 3-11 Normalized centering force versus normalized displacement, Case (3), and Case (5) (Chapter 4); $Q_L = 15$, $Q_0 = 10$.

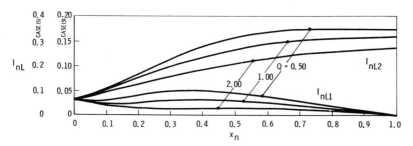

Figure 3-12 Normalized coil currents versus normalized displacement, Case (3), and Case (5) (Chapter 4); $Q_L = 15$, $Q_0 = 10$.

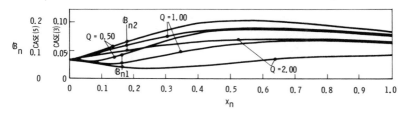

Figure 3-13 Normalized flux densities versus normalized displacement, Case (3), and Case (5) (Chapter 4); $Q_L = 15$, $Q_0 = 10$.

under this circumstance. The ratio of stiffnesses for voltage-source operation with parallel tuning and current-source operation with series tuning, Cases (3) and (2), is

$$\left. \frac{\dot{F}_{Vp}}{\dot{F}_{Is}} \right|_{x \to 0} = \frac{V^2}{R^2 I^2} \frac{1}{4(Q_L^2 + 1)}. \tag{3-21}$$

The conditions for stability for Case (3) are the same as derived in Section 2-4 for Case (1), since the numerators of Equations (3-18) and (3-19) are the same as previously treated and must be negative. However, the conditions for maximum stiffnesses are different. Differentiating Equation (3-19) with respect to Q and equating to zero shows that maximum stiffness with the block centered occurs for $Q = 1$ for this case also, for all values of Q_L and Q_0. The corresponding normalized peak stiffnesses are

$$\dot{F}_{nm} = -\frac{Q_0 - 2}{2(Q_L^2 + 1)}. \tag{3-22}$$

Whereas the definition of Q for parallel tuning, Cases (3) and (4), is the same as for series tuning, Cases (1) and (2), in terms of the parameters involved, it cannot be interpreted in terms of power points. Some plots of $\dot{F}_{n0}$ from Equation (3-19) as a function of Q are shown in Figure 3-14 for $Q_L/Q_0 = \frac{3}{2}$, which means $Q_\ell/Q_L = \frac{1}{3}$, or 33.3 percent leakage flux. These plots can be obtained by dividing the ordinates of Figure 3-9 by $4(Q_L^2 + 1)$ for selected values of $Q_L > Q_0$.

Substitution of the expressions for the various admittances and impedances into the second form of Equation (3-17) gives for Case (4)

$$F_{Ip} = \frac{N^2 \mu A}{g_0^2} \frac{I^2}{4\omega^2 C^2 R^2} \frac{4(Q^2 - Q_0 Q + 1)x_n}{\{[Q_0 - (Q_0 - Q)(1 - x_n)]^2 + (1 - x_n)^2\}} , \tag{3-23}$$
$$\times \{[Q_0 - (Q_0 - Q)(1 + x_n)]^2 + (1 + x_n)^2\}$$

a familiar form, in fact identical to Equation (2-7) in form. However, the apparent normalizing factor is not the pull on one side when the block is centered; further, since the tuning is accomplished by setting capacitance C, this normalizing factor cannot be kept fixed as Q is changed unless I also is changed. The pull on one side, when the block is centered, is

$$F_I = \frac{N^2 I^2 \mu A}{g_0^2} \frac{(Q_L - Q)^2}{Q^2 + 1}, \tag{3-24}$$

but it changes with Q_L and Q. If the same normalizing factor is used as for

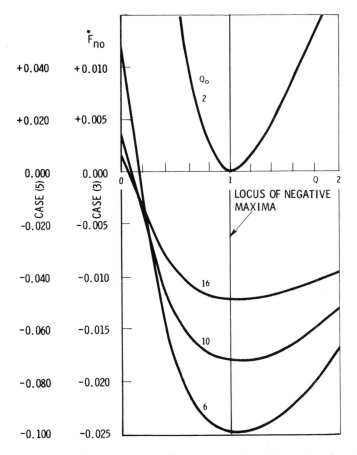

Figure 3-14 Normalized initial stiffness versus Q for various values of Q_0, Case (3), and Case (5) (Chapter 4); $Q_L/Q_0 = \frac{3}{2}$.

series tuning, Case (2), Equation (3-3), to give a fixed base and have direct comparison *for equal current sources*, the normalized version of Equation (3-23) becomes

$$F_n = \frac{16(Q_L - Q)^2(Q^2 - Q_0Q + 1)x_n}{\{[Q_0 - (Q_0 - Q)(1 - x_n)]^2 + (1 - x_n)^2\} \times \{[Q_0 - (Q_0 - Q)(1 + x_n)]^2 + (1 + x_n)^2\}}. \tag{3-25}$$

The normalized coil currents, on the same base as for Case (2), can be obtained from

$$I_{L1} = \frac{I}{2} \frac{2(Q_L - Q)(1 - x_n)}{\sqrt{[Q_0 - (Q_0 - Q)(1 - x_n)]^2 + (1 - x_n)^2}}$$

$$= \frac{I}{2} \frac{2(Q_L - Q)}{\sqrt{1 + \left(Q + \dfrac{Q_0 x_n}{1 - x_n}\right)^2}} = \frac{I}{2} I_{nL1}$$

and

$$I_{L2} = \frac{I}{2} \frac{2(Q_L - Q)(1 + x_n)}{\sqrt{[Q_0 - (Q_0 - Q)(1 + x_n)]^2 + (1 + x_n)^2}}$$

$$= \frac{I}{2} \frac{2(Q_L - Q)}{\sqrt{1 + \left(Q - \dfrac{Q_0 x_n}{1 + x_n}\right)^2}} = \frac{I}{2} I_{nL2},$$

and the relation between normalized currents and normalized flux densities is as derived for the previous cases. Normalized curves for F_n, I_{nL}, and $\mathscr{B}_n$ versus x_n are plotted in Figures 3-15 through 3-21, for various combinations of Q_L, Q_0, and Q. By use of corresponding Figures 2-3 through 2-11 for Case (1), the

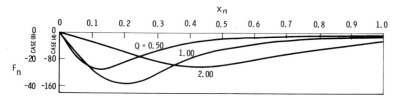

Figure 3-15 Normalized centering force versus normalized displacement, Case (4), and Case (8) (Chapter 4); $Q_L = 10$, $Q_0 = 6$.

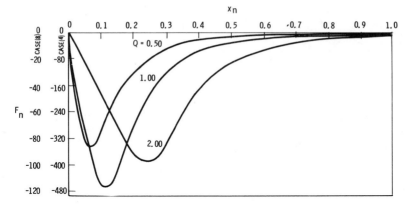

Figure 3-16 Normalized centering force versus normalized displacement, Case (4), and Case (8) (Chapter 4); $Q_L = 15$, $Q_0 = 10$.

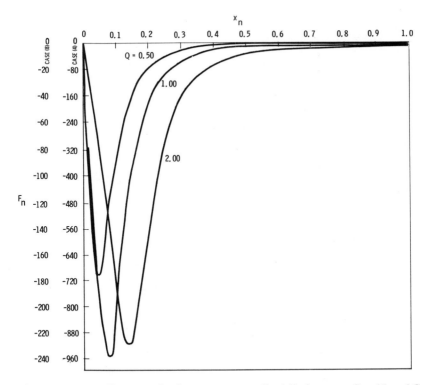

Figure 3-17 Normalized centering force versus normalized displacement, Case (4), and Case (8) (Chapter 4); $Q_L = 20$, $Q_0 = 16$.

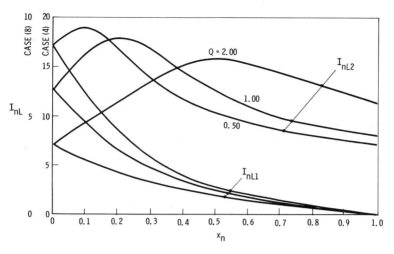

Figure 3-18 Normalized coil currents versus normalized displacement, Case (4), and Case (8) (Chapter 4); $Q_L = 10$, $Q_0 = 6$.

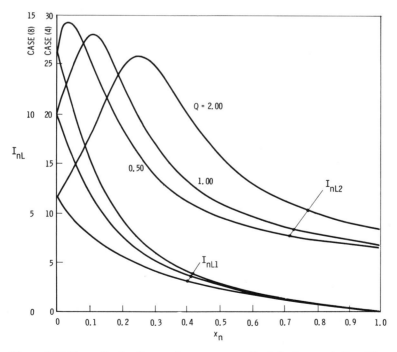

Figure 3-19 Normalized coil currents versus normalized displacement, Case (4), and Case (8) (Chapter 4); $Q_L = 15$, $Q_0 = 10$.

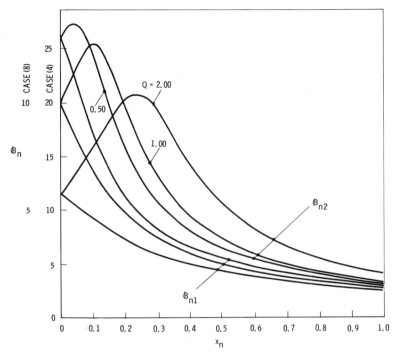

Figure 3-20 Normalized flux densities versus normalized displacement, Case (4), and Case (8) (Chapter 4); $Q_L = 15, Q_0 = 10$.

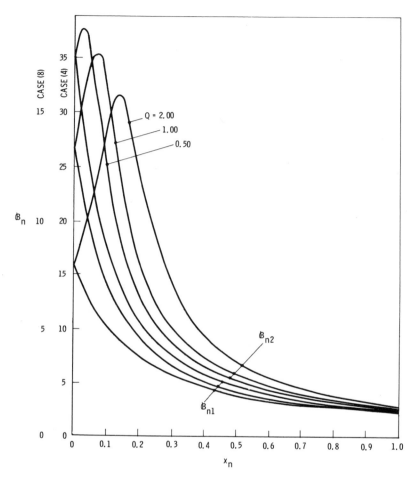

Figure 3-21 Normalized flux densities versus normalized displacement, Case (4), and Case (8) (Chapter 4); $Q_L = 20$, $Q_0 = 16$.

force curves for Case (4) can be obtained by multiplying the ordinates by $4(Q_L - Q)^2$, and the current and flux density curves for Case (4) can be obtained by multiplying the ordinates by $2(Q_L - Q)$ for selected values of $Q_L > Q_0$.

The normalized forces for operation with current source and with voltage source are related by

$$\frac{F_{nI}}{F_{nV}} = \frac{(Z_1 + Z_2)^2}{R^2} = \frac{(Y_1 + Y_2)^2}{(Y_1 Y_2)^2 R^2} \tag{3-26}$$

if $V = RI$, and the actual average forces are related by

$$\frac{F_{Ip}}{F_{Vp}} = \frac{I^2}{V^2}(Z_1 + Z_2)^2. \tag{3-27}$$

The impedances and admittances here of course are not the same as for Equations (3-4) and (3-5) that apply to series tuning, Cases (1) and (2), but are as defined for parallel tuning following Equation (3-15).

The stiffness at zero displacement for Case (4) is

$$\dot{F}_{Ip}\bigg|_{x \to 0} = \frac{N^2 I^2 \mu A}{4g_0^3} \frac{16(Q_L - Q)^2(Q^2 - Q_0 Q + 1)}{(Q^2 + 1)^2} = \frac{N^2 I^2 \mu A}{4g_0^3} \dot{F}_{n0}. \tag{3-28}$$

Plots of $\dot{F}_{n0}$ from Equation (3-28), Figure 3-22, can be obtained by multiplying the ordinates of Figure 2-5 also by $4(Q_L - Q)^2$, for $Q_L > Q_0$. Here $Q_L/Q_0 = \frac{3}{2}$ is used throughout, as for Figure 3-14. The algebraic relation for the locations of the stiffness peaks in Figure 3-22 is too complicated to be practical, being fifth order in Q. The conditions for stability are the same as derived in Section 2-4. The stiffnesses *for equal current sources*, for parallel and series tuning, Cases (4) and (2), are related by

$$\frac{\dot{F}_{Ip}}{\dot{F}_{Is}}\bigg|_{x \to 0} = \frac{4(Q_L - Q)^2}{Q^2 + 1}. \tag{3-29}$$

Evidently, the parallel arrangement, Case (4), is then much the stiffer. The stiffnesses with parallel tuning for current-source and for voltage-source operation, Cases (4) and (3), are related by

$$\frac{\dot{F}_{Ip}}{\dot{F}_{Vp}}\bigg|_{x \to 0} = 16 \frac{R^2 I^2}{V^2} \frac{(Q_L - Q)^2(Q_L^2 + 1)}{Q^2 + 1}. \tag{3-30}$$

Hence if $V = RI$, the stiffness with block centered is many times as much for

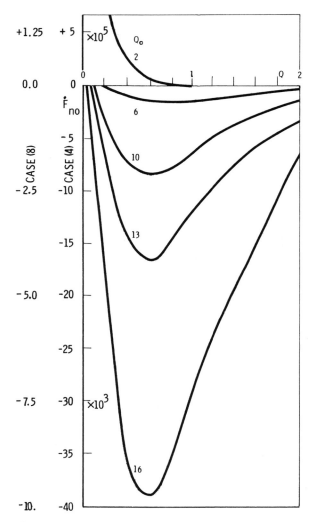

Figure 3-22 Normalized initial stiffness versus Q for various values of Q_0, Case (4) and Case (8) (Chapter 4); $Q_L/Q_0 = \frac{3}{2}$.

current-source operation as for voltage-source operation, with parallel tuning, Cases (4) and (3). The comparison of stiffnesses for current-source operation with parallel tuning and for voltage-source operation with series tuning, Cases (4) and (1), is

$$\left.\frac{\dot{F}_{Ip}}{\dot{F}_{Vs}}\right|_{x\to 0} = \frac{R^2 I^2}{V^2} 4(Q_L - Q)^2. \tag{3-31}$$

The quiescent coil current can become very large for large Q_L and small Q with current-source operation, Case (4):

$$I_{RL} = \frac{I(Q_L - Q)}{\sqrt{Q^2 + 1}}. \tag{3-32}$$

For sharp tuning, the current to the capacitor is nearly the same as the coil current. The voltage across the parallel combination is

$$V_p = \frac{V}{2} = 2RI(Q_L - Q)\sqrt{\frac{Q_L^2 + 1}{Q^2 + 1}}. \tag{3-33}$$

As a gap opens, the coil current and voltage on that side increase.

As for Cases (1) and (2), the relations involving Cases (3) and (4) are based on the assumption of pure voltage and current sources. However, if the equivalent series impedance of the voltage source is not zero but very small or if the equivalent shunt impedance of the current source is not infinite but very large, results are not greatly affected if the displacement of the suspended member is relatively small, because the impedances of the two parallel combinations change in opposite directions and the net impedance of the two parallel combinations in series then does not change greatly.

3-4 Idealized Parallel Tuning, Cases (3a) and (4a)

An idealized condition[17,38] having some possible interest is the use of a voltage source with parallel tuning and a resistanceless coil having no leakage inductance, Case (3a). The "Q" concept then becomes useless, but the result can be expressed in terms of inductance and capacitance:

$$F_{Vp} = -\frac{N^2 V^2 \mu A}{4\omega^2 L_0^2 g_0^2}\left(\frac{x_n}{2\omega^2 L_0 C - 1}\right). \tag{3-34}$$

This relation is perfectly linear, the stiffness being constant at

$$\dot{F}_{Vp} = -\frac{N^2 V^2 \mu A}{4\omega^2 L_0^2 g_0^3} \left(\frac{1}{2\omega^2 L_0 C - 1}\right). \tag{3-35}$$

This arrangement has no stability problem, provided only that

$$2\omega^2 L_0 C - 1 > 0, \tag{3-36}$$

which means that the resonant condition with $x = 0$ should not be approached. That situation would lead also to large coil and capacitor currents. If shunt conductance G is added, as shown in Figure 3-10, but leakage inductance L_l and coil resistance R still are zero, the linearity still exists, in a situation one step closer to practicality. Then

$$\begin{aligned} F_{Vp} &= -\frac{N^2 V^2 \mu A}{4\omega^2 L_0^2 g_0^2} \frac{\left(\omega C - \dfrac{1}{2\omega L_0}\right) x_n}{2\omega L_0 \left[G^2 + \left(\omega C - \dfrac{1}{2\omega L_0}\right)^2\right]} \\ &= -\frac{N^2 V^2 \mu A G^2}{4g_0^2} \left(\frac{4D_{L0}^3 D x_n}{D^2 + 1}\right) = -\frac{N^2 V^2 \mu A G^2}{4g_0^2} F_n, \end{aligned} \tag{3-37}$$

in which

$$D_{L0} = \frac{1}{2\omega L_0 G}$$

and

$$D = \frac{\omega C}{G} - \frac{1}{2\omega L_0 G} = D_c - D_{L0}.$$

The stiffness is

$$\dot{F}_{Vp} = -\frac{N^2 V^2 \mu A G^2}{4g_0^3} \left(\frac{4D_{L0}^3 D}{D^2 + 1}\right) = -\frac{N^2 V^2 \mu A G^2}{4g_0^3} \dot{F}_n. \tag{3-38}$$

Relation (3-36) gives the stability criterion for this arrangement also.

An actual coil can be represented as a conductance in parallel with an inductance, and both equivalent parameters are fixed at fixed frequency but depend on actual coil resistance and inductance. For a coil having high Q_L the equivalent conductance is quite small, and the equivalent inductance is nearly equal to the actual inductance; but if actual inductance varies, the equivalent conductance varies. Hence, to achieve a good approximation to a linear suspension, leakage inductances should be kept as small as possible, and the effect of variation of the equivalent parallel conductance of the coil should be smothered by relatively large actual parallel conductance. The procedure unfortunately tends to give relatively low stiffness. Use of a current source with the arrangement of Figure 3-10, Case (4a), does not give a linear result for zero resistance or for zero resistance and zero conductance, with or without zero leakage inductance.

3-5 Position Signals and Adjustment

The circuit of Figure 3-10 does not of itself give a bridge circuit arrangement whereby the position of the block may be monitored. A bridge circuit might be established by placing an auxiliary branch across the source and monitoring the voltage between a mid-tap on that branch and the junction of the two parallel combinations. Usually, however, a differential transformer is used to compare the coil currents, the effect of which is equivalent to a slight increase in the leakage inductances of the coils. In terms of a transformation constant k_v, the output voltage signal of this transformer can be written as

$$V_B = k_v R(I_{L2} - I_{L1}) = k_v R\left(\frac{V_2}{Z_{L2}} - \frac{V_1}{Z_{L1}}\right)$$

$$= k_v R V\left[\frac{1}{Z_{L2}(1 + Y_2 Z_1)} - \frac{1}{Z_{L1}(1 + Y_1 Z_2)}\right] \qquad (3\text{-}39)$$

or, if the expression is evaluated for very small displacements x_n by insertion of the various expressions for admittance and dropping terms in x_n^2, as

$$|V_n| = \left|\frac{V_B}{V}\right| \approx \frac{k_V Q_0 x_n}{2\sqrt{(Q_L^2 + 1)(Q^2 + 1)}}, \qquad (3\text{-}40)$$

with voltage-source operation, Case (3), and no shunt conductance. For current-source operation, Case (4),

$$V_B = k_I R(I_{L2} - I_{L1}) = k_I R I\left(\frac{Y_{L2}}{Y_2} - \frac{Y_{L1}}{Y_1}\right), \qquad (3\text{-}41)$$

or

$$|V_n| = \left|\frac{V_B}{RI}\right| \approx \frac{2k_I Q_0(Q_L - Q)x_n}{Q^2 + 1} \qquad (3\text{-}42)$$

for very small displacements.

The condition for adjusting the position of the block by capacitive trimming is

$$\left|\frac{I_{L1}}{I_{L2}}\right| = \frac{g_0 - x_0}{g_0 + x_0} = \left|\frac{Y_{L1} Y_2}{Y_1 Y_{L2}}\right|$$

$$= \sqrt{\frac{\left[1 - 2\omega^2 C(1 + c_n)\left(L - L_0\dfrac{x_{n0}}{1 + x_{n0}}\right)\right]^2 + [2\omega CR(1 + c_n)(1 + r_n)]^2}{\left[1 - 2\omega^2 C(1 - c_n)\left(L + L_0\dfrac{x_{n0}}{1 - x_{n0}}\right)\right]^2 + [2\omega CR(1 - c_n)(1 - r_n)]^2}}$$

(3-43)

or

$$\left(\frac{1 - x_{n0}}{1 + x_{n0}}\right)^2 = \left(\frac{1 - x_{n0}}{1 + x_{n0}}\right)^2 \frac{\{(1 + c_n)[Q_0 - (Q_L - Q_0)(1 + x_{n0})] + (Q_L - Q)(1 + x_{n0})\}^2 + [(1 + c_n)(1 + r_n)(1 + x_{n0})]^2}{\{(1 - c_n)[Q_0 - (Q_L + Q_0)(1 - x_{n0})] + (Q_L - Q)(1 - x_{n0})\}^2 + [(1 - c_n)(1 - r_n)(1 - x_{n0})]^2},$$

(3-44)

solution of which for small displacements gives

$$x_{n0} \approx -[r_n + (Q_L Q + 1)c_n]$$

(3-45)

or

$$c_n \approx -\frac{r_n + x_{n0}}{Q_L Q + 1}.$$

(3-46)

3-6 Winding Capacitance

In all of the derivations winding capacitance of the coils has been ignored. For the series-tuning connection, Cases (1) and (2), such capacitance could influence the behavior of the suspension at frequencies high enough to render the capacitive shunting susceptance large enough to be significant. However, for the parallel tuning connection, Cases (3) and (4), the winding capacitance can as a first approximation be considered to be a part of the tuning capacitance, and its first effect therefore would be merely to reduce the amount of added parallel capacitance C needed to establish the desired operating condition. An interesting speculation is that at sufficiently high frequency the coils alone might provide a stable suspension without the use of added tuning capacitance.

3-7 Summary

Self-stabilizing or passive suspensions can be developed by means of tuned circuitry with either voltage or current sources and with either series or parallel tuning. The behaviors in terms of restoring force versus displacement, the stiffness, the coil currents and voltages, and means for detecting and adjusting the position of the suspended member are derived under Cases (1) through (4) in Chapters 2 and 3, with intercomparisons. For purposes of classification, the connections are identified as follows:

Case (1). Coils and capacitor for each pole pair in series across a common voltage source, Figure 2-1.

Case (2). Coils and capacitor for each pole pair in series across a common current source, Figure 2-1.

Case (3). Coils for each pole pair in series, and in parallel with a capacitor; parallel combinations in series with a voltage source, Figure 3-10.

Case (4). Coils for each pole pair in series, and in parallel with a capacitor; parallel combinations in series with a current source, Figure 3-10.

In the following chapter, four other cases are considered, in which the four coils of Figures 2-1 or 3-10 form the arms of an impedance bridge.

4 BRIDGE-CIRCUIT CONNECTIONS FOR SINGLE-AXIS MAGNETIC SUSPENSIONS

In these connections, the bridge circuit[39]* is formed for actual operational purposes and is not merely an incidental derivation as a means of indicating the position of the suspended member,[1,2]† though that purpose is served also. The capacitance in the detector position does not accomplish true series or parallel tuning, though the result is largely the same. The advantage is that only one capacitor is required instead of two. The total circuit can be series or parallel tuned by using a capacitor in association with the source.

4-1 Voltage Source, No Series Impedance, Case (5)

In Figure 4-1, the coils at one end of the block, Figure 2-1, are supposed to be in opposite branches of the bridge circuit, and the coils at the other end in the other opposite branches with $L_2 = L_3$ and $L_1 = L_4$. No impedance is supposed to be in the source branch, as indicated in Figure 4-1 by capacitance $2C$ shown dotted, and the trimming capacitances c are supposed to be absent. Strictly, then, the circuit is not tuned, because the capacitance in the detector branch has no influence at balance. Nevertheless, for comparison with the behavior of circuits analyzed in preceding chapters, definition of Q as

$$Q = \frac{\omega L}{R} - \frac{1}{2\omega CR}$$

still is useful, though it again is not indicative of power point and, further, now is not indicative of sharpness of tuning. Solution of the bridge circuit for its currents in terms of source voltage and impedances gives

$$
\begin{aligned}
F_{VB} = 4F_{Vp} &= \frac{N^2 V^2 \mu A}{4R^2 g_0^2} \frac{4(Q^2 - Q_0 Q + 1)x_n}{\{Q_L + Q - [(Q_L - Q_0) - (Q_0 - Q)]x_n^2\}^2 + \{Q_L Q - 1 + [(Q_L - Q_0)(Q_0 - Q) + 1]x_n^2\}^2} \\
&= \frac{N^2 V^2 \mu A}{4R^2 g_0^2} F_n,
\end{aligned}
$$

(4-1)

which is the same as Equation (3-18) for voltage-source operation with parallel tuning, Case (3), except for a factor of 4. As indicated in Section 3-3 in

---

*Suggested by P. J. Gilinson, Jr., in October 1969, patent applied for and pending.
† Adaptations of the original "mesh" circuit are discussed in Sections 7–5 and 9–1.

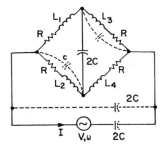

Figure 4-1 Bridge-circuit connections for single-axis magnetic suspension, Cases (5) through (8).

discussion of Equation (3-18), the factor of 4 could be retrieved in it by using source voltage $2V$, which would give quiescent voltage V across each coil pair.

The conditions for stability are the same as derived in Section 2-4 for voltage-source operation with series tuning, Case (1). The conditions with the block centered are independent of the setting of C and hence independent of Q, the quiescent current in a coil being

$$I_0 = \frac{V}{2R\sqrt{Q_L^2 + 1}}.$$

The stiffness at zero displacement is

$$\dot{F}_{VB}\big|_{x \to 0} = \frac{N^2 V^2 \mu A}{4R^2 g_0^3} \frac{4(Q^2 - Q_0 Q + 1)}{(Q_L^2 + 1)(Q^2 + 1)} = \frac{N^2 V^2 \mu A}{4R^2 g_0^3} \dot{F}_{n0}, \tag{4-2}$$

which is the same as Equation (3-19) for voltage-source operation with parallel tuning, Case (3), except for a factor of 4. Hence the stiffness for operation with this bridge circuit, Case (5), compared with operation with parallel tuning, Case (3), *for equal voltage sources* is

$$\frac{\dot{F}_{VB}}{\dot{F}_{Vp}}\bigg|_{x \to 0} = 4; \tag{4-3}$$

the stiffness compared with current-source operation with parallel tuning, Case (4), is

$$\frac{\dot{F}_{VB}}{\dot{F}_{Ip}}\bigg|_{x \to 0} = \frac{V^2}{R^2 I^2} \frac{Q^2 + 1}{4(Q_L^2 + 1)(Q_L - Q)^2}, \tag{4-4}$$

which is 4 times the reciprocal of Equation (3-30); the stiffness compared with voltage-source operation with series tuning, Case (1), *for equal voltage sources* is

$$\frac{\dot{F}_{VB}}{\dot{F}_{Vs}}\Bigg|_{x\to 0} = \frac{Q^2 + 1}{Q_L^2 + 1},\tag{4-5}$$

which is the same as Equation (3-20) except for lacking a factor of 4 in the denominator; the stiffness compared with current-source operation with series tuning, Case (2), is

$$\frac{\dot{F}_{VB}}{\dot{F}_{Is}}\Bigg|_{x\to 0} = \frac{V^2}{R^2 I^2}\frac{1}{Q_L^2 + 1},\tag{4-6}$$

which is the same as Equation (3-21) except for lacking a factor of 4 in the denominator. The plots of normalized force and stiffness, Figures 3-11 and 3-14 apply to this connection by use of ordinate scales expanded by a factor of 4.

4-2 Current Source, No Series Impedance, Case (6)

Impedance in series with a current source does not affect the behavior of the suspension; it merely determines the voltage that appears across the source as a function of position of the block. With a current source in place of the voltage source of Figure 4-1, the expression for average force is the same as Equations (3-2) and (3-3) for current-source operation with series tuning, Case (2), and Figures 3-2 through 3-4 apply directly; $F_{IB} = F_{Is}$.

The stiffness relations are given by Equations (3-7), (3-8), and (3-10), and Figure 3-9 applies directly. The stiffness compared with the stiffnesses for other cases are the same as for Case (2), Equations (3-9), (3-21), (3-29), and (4-6).

4-3 Voltage Source, Tuned Bridge Circuit, Case (7)

If now the capacitance $2C$, shown dotted in Figure 4-1, is inserted in series with voltage source V and, being the same as the capacitance in the detector branch, satisfies the definition of Q, then in that relationship Q has its original significance with respect to sharpness of tuning of the circuit as a whole when the block is centered and with respect to the power-point setting for the circuit as a whole in the quiescent state. Then solution of the bridge circuit for its currents in terms of source voltage and impedances to obtain average force F_{VBT} is the same as Equation (2-7) for F_{Vs}, Case (1). The conditions for stability and the stiffness relations are therefore the same as for voltage-source operation with series tuning, Case (1), as given in Sections 2-4 and 2-6, and the plots of Figures 2-3 through 2-5 apply directly. The stiffness is the same as plotted on Figures 2-12 and 2-13 for Case (1) and, as related to stiffness resulting from other circuit connections, is the same as given in Equations (3-9), (3-20), (3-31), and (4-5).

4-4 Current Source, Tuned Bridge Circuit, Case (8)
The tuning can be accomplished also by placing capacitance $2C$ across the bridge circuit in parallel with current source I, Figure 4-1. The transformation between voltage and current source is given by

$$V = \frac{I}{2\omega C} = \frac{IR}{2\omega CR} = IRQ_c = IR(Q_L - Q).$$

Substitution for V in terms of I in Equation (2-7) gives $F_{IBT} = F_{Ip}/4$, or, in other words, the force for the tuned bridge circuit with a current source, Case (8), is $\frac{1}{4}$ as much as the force for parallel tuning with a current source, Equation (3-23), Case (4), *for equal current sources*. Likewise $\dot{F}_{IBT} = \dot{F}_{Ip}/4$ can be obtained from Equation (3-28), and the relations of the stiffness for Case (8) to the stiffnesses for Case (2) = Case (6), Case (3), or Case (1) = Case (7) can be obtained by dividing by 4 Equations (3-29), (3-30), and (3-31); the stiffness for Case (5) is related to the stiffness for Case (8) by 4 times Equation (4-4).

4-5 Branch Currents and Voltages
The coil currents for voltage-source operation with the tuned bridge circuit, Case (7), are the same as the coil currents for voltage-source operation with series tuning, Case (1), *when the source voltages are the same*. Hence the coil voltages for those two cases are the same *when the source voltages are equal*. The coil currents for voltage-source operation without tuning, Case (5), are twice as large as the coil currents for voltage-source operation with parallel tuning, Case (3). Hence, *for equal source voltages*, the coil voltages for the bridge circuit, Case (5), are twice as large as for the parallel-tuned circuit, Case (3). The variation of coil currents with respect to displacement of the block are quite different for the tuned bridge circuit and for the untuned bridge circuit, as are the coil-current variations for series tuning and for parallel tuning, with a voltage source.

Likewise, the coil currents for current-source operation of the bridge circuit, Case (6), are the same as the coil currents for current-source operation with series tuning, Case (2), *for equal source currents*, and the coil voltages for the two cases are, therefore, the same. The coil currents for current-source operation with the tuned bridge circuit, Case (8), are only half as large as the coil currents for current-source operation with parallel tuning, Case (4), *for equal source currents*. Hence for equal source currents the coil voltages for the bridge circuit, Case (8), are only half as large as for the parallel-tuned circuit, Case (4). The coil-current variations with respect to block displacement for the tuned bridge circuit and the untuned bridge circuit are quite different, as are the coil-current variations for series tuning and parallel tuning, with a current source.

For the bridge circuit, the voltage across the capacitor in the detector branch is not the same or essentially the same as the coil voltage. It is proportional to the difference of the currents at the two ends of the block and is zero when the block is centered. When the bridge circuit is tuned for voltage-source operation, Case (7), the voltage across the tuning capacitor is proportional to the sum of the currents at the two ends of the block and at zero displacement is the same as the voltage across the tuning capacitor for voltage-source operation with series tuning, Case (1), *for equal voltage sources*. When the bridge circuit is tuned for current-source operation, Case (8), the voltage across the tuning capacitor is the same as the voltage across the bridge circuit and at zero displacement is half the voltage across the tuning capacitor for current-source operation with parallel tuning, Case (4), *for equal current sources*. For any of these bridge-circuit conditions, the voltages across the capacitor do not become as large when the block is displaced as do the voltages across the capacitors for series or parallel tuning with equal voltage sources or with equal current sources. These questions of voltage and current limitations are discussed more generally in connection with other limitations in Chapter 5.

4-6 Position Signals

For the bridge circuit, the voltage across the detector branch is a natural indication of the position of the block. In general,

$$V_B = \frac{I_2 - I_1}{j2\omega C}. \tag{4-7}$$

For voltage-source operation without tuning, Case (5), and for very small displacements,

$$|V_n| = \left|\frac{V_B}{V}\right| \approx \frac{Q_0(Q_L - Q)x_n}{\sqrt{(Q_L^2 + 1)(Q^2 + 1)}}, \tag{4-8}$$

which is twice Equation (3-40) for the parallel tuned circuit, Case (3), if $k_V = Q_L - Q$. For current-source operation, Case (6), and for very small displacements,

$$|V_n| = \left| \frac{V_B}{RI} \right| \approx \frac{Q_0(Q_L - Q)x_n}{\sqrt{Q^2 + 1}}, \tag{4-9}$$

which is half as sensitive as the signal for the series-tuned circuit, Case (2), Equation (3-14), *with equal current sources*. For voltage-source operation with a tuned bridge circuit, Case (7), and for very small displacements,

$$|V_n| = \left| \frac{V_{BT}}{V} \right| \approx \frac{Q_0(Q_L - Q)x_n}{Q^2 + 1}, \tag{4-10}$$

which is half as sensitive as the series-tuned circuit, Case (1), Equation (2-22), *with equal voltage sources*. For current-source operation with a tuned bridge circuit, Case (8), and for very small displacements,

$$\frac{V_{BT}}{RI} = \frac{(Q_L - Q)^2 Q_0 x_n}{Q^2 + 1}, \tag{4-11}$$

which is half of Equation (3-42) for the parallel-tuned circuit, Case (4), if $k_I = Q_L - Q$.

4-7 Initial Positioning by Trimming Capacitance

The initial force center position of the block can be adjusted by use of trimming capacitance c as shown shunted across branches 2 and 3 in Figure 4-1, but to achieve the desired direction of initial displacement the capacitances may need to be shunted around branches 1 and 4. Other means of adjustment could be used, of course. If, as is quite likely and has been assumed for the series and parallel tuned cases, the coil resistances are not quite alike, all four resistances would have to be assumed to be different in this case, a situation that leads to a considerable algebraic maze in solving for the relation among these resistances, the trimming capacitances, and the quiescent displacement. But if the resistances are taken to be equal and the windings alike, the relation between quiescent displacement and trimming capacitance for very small displacements is simply

$$x_{n0} \approx \frac{(Q_L Q + 1)c_n}{2[Q(Q_0 - Q) - 1]} \tag{4-12}$$

or

$$c_n \approx \frac{2[Q(Q_0 - Q) - 1]x_{n0}}{Q_L Q + 1}, \tag{4-13}$$

derived by letting

$$\left|\frac{I_1}{I_2'}\right|^2 = \left|\frac{I_4}{I_3'}\right|^2 = \frac{(1 - x_{n0})^2}{(1 + x_{n0})^2}, \tag{4-14}$$

which applies for all four source conditions, because the branch-current *ratios* are not affected by those conditions. Here I_2' and I_3' represent the respective currents in coils 2 and 3, Figure 4-1, which are at end 2 of the block. If c is incorporated into branches 1 and 4, the relative signs of c_n and x_n reverse. In actual practice the four coils are not exactly equal, and the foregoing relations can be taken as bases for a first trial at experimental trimming.

4-8 Summary
The bridge-circuit connections of the coils of Figure 4-1, Cases (5) through (8), can give the same force-displacement relations as for series and parallel tuning, Figures 2-1 and 3-10, Cases (1) through (4), and except when the bridge circuit as a whole is tuned, Cases (7) and (8), only one instead of two capacitor units is required. The force comparisons are

F_{VB} [Case (5)] $= 4F_{Vp}$ [Case (3)], equal V,
F_{IB} [Case (6)] $= F_{Is}$ [Case (2)], equal I,
F_{VBT} [Case (7)] $= F_{Vs}$ [Case (1)], equal V,
and
F_{IBT} [Case (8)] $= F_{Ip}/4$ [Case (4)], equal I.

Hence, while eight different circuit connections have been illustrated in Chapters 2, 3, and 4, only four of them give different performances of the suspension.

5 COMPARISON OF PERFORMANCES FOR VARIOUS OPERATING CONDITIONS

To this point the performance of the suspension has been based on the application of a sinusoidal voltage or current source having constant amplitude. In this chapter the performances resulting from the various circuit connections with such excitation are compared. However, other bases of comparison sometimes are preferable, depending on the application. For example, a fixed limit may be placed on the coil current, the coil voltage, the flux density, the maximum force, or the stiffness. Further, the excitation may be square wave or other nonsinusoidal form, or may consist of pulses or pulse trains. Performance comparisons are given for these other operating conditions also.

5-1 Constant-Amplitude Sinusoidal Voltage or Current Source

This type of operation is analyzed for various circuit conditions designated as Cases (1) through (8) in Chapters 2, 3, and 4, and performance curves of normalized restoring forces, currents, and flux densities versus normalized displacements, and normalized stiffness versus quality factor Q are shown. Of these cases only four really are different, aside from factors of 4 relating Cases (3) and (5) and Cases (8) and (4). Comparisons of normalized maximum stiffnesses, maximum forces, coil currents for maximum forces, maximum flux densities, flux densities at maximum forces, and the normalized displacements at which they occur are shown in Table 5-1. Normalized coil current corresponding to maximum restoring force is tabulated, rather than normalized maximum coil current, because the maximum coil current occurs either quite close to or substantially beyond the displacement corresponding to maximum restoring force. The heating effect of current at displacement beyond the displacement for maximum restoring force is of little concern, because it exists only during erection. Maximum normalized coil-2 voltages, maximum capacitor voltages, and the displacements at which they occur are not shown, but the normalized quiescent coil-2 voltages are shown. In general the maximum voltage across the associated capacitor is less than or equal to the maximum voltage across coils 2. Derivation of the exact maxima for coil and capacitor voltages and the displacements at which they occur is very complicated algebraically in some cases. However, in those cases approximations can be made that are adequate for practical purposes, since generous factors of safety ordinarily are used to guard against insulation breakdown.

Table 5-1. Equal Voltage or Current Sources, or $V = RI$

Case	Q_L	Q_o	Q	$\dot{F}_{no}$	F_{nm}	x_{nF}	$\mathcal{B}_{n2F}$	I_{nLF}	$\mathcal{B}_{n2m}$	x_{nB}	V_{nRLO}
1, 7	10	6	0.5	− 4.4	− 0.31	0.12		0.99	0.932	0.060	8.99
2, 6	10	6	0.5	− 7.6	− 1.16	0.31	1.687	2.23	1.72	0.380	10.05
3, 5*	10	6	0.5	− 0.019	− 0.0041	0.47	0.112	0.160	0.12	†	0.50
4, 8*	10	6	0.5	− 1600.	− 109.	0.12		18.80	17.7	0.060	171.
1, 7	10	6	1.0	− 4.0	− 0.48	0.19		1.00	0.850	0.154	7.10
2, 6	10	6	1.0	− 8.3	− 1.66	0.34	1.537	2.05	1.55	0.380	10.05
3, 5*	10	6	1.0	− 0.025	− 0.0059	0.52	0.098	0.15	0.099	†	0.50
4, 8*	10	6	1.0	− 1300.	− 153.	0.19		18.0	15.29	0.154	128.
1, 7	10	6	2.0	− 1.1	− 0.40	0.44		0.99	0.687	0.412	4.49
2, 6	10	6	2.0	− 5.6	− 1.48	0.45	1.314	1.90	1.32	0.400‡	10.05
3, 5*	10	6	2.0	− 0.017	− 0.0058	0.69	0.082	0.14	0.082	†	0.50
4, 8*	10	6	2.0	− 290.	− 102.	0.44		15.80	10.99	0.412	72.
1, 7	15	10	0.5	− 9.5	− 0.41	0.07		0.99	0.955	0.041	13.45
2, 6	15	10	0.5	− 14.8	− 1.81	0.24	2.046	2.55	2.10	0.31	15.04
3, 5*	15	10	0.5	− 0.015	− 0.0031	0.39	0.100	0.14	0.109	†	0.50
4, 8*	15	10	0.5	− 8100.	− 349.	0.07		28.70	27.5	0.041	390.
1, 7	15	10	1.0	− 8.0	− 0.61	0.11		1.00	0.906	0.098	10.62
2, 6	15	10	1.0	− 15.2	− 2.41	0.25	1.830	2.30	1.86	0.32	15.04
3, 5*	15	10	1.0	− 0.018	− 0.0043	0.41	0.088	0.12	0.091	†	0.50
4, 8*	15	10	1.0	− 6300.	− 473.	0.11		28.0	25.4	0.098	298.
1, 7	15	10	2.0	− 2.4	− 0.58	0.24		1.00	0.806	0.231	6.73
2, 6	15	10	2.0	− 12.5	− 2.19	0.32	1.558	2.05	1.56	0.36	15.04
3, 5*	15	10	2.0	− 0.013	− 0.0042	0.54	0.073	0.108	0.074	†	0.50
4, 8*	15	10	2.0	− 1620.	− 395.	0.24		26.0	20.96	0.231	175.

1, 7	20	16	0.5	— 17.3	— 0.461	0.046	2.434	0.99	0.971	0.028	17.95
2, 6	20	16	0.5	— 26.7	— 2.50	0.18	0.112	2.79	2.55	0.24	20.02
3, 5*	20	16	0.5	0.010	— 0.0029	0.38		0.153	0.123	†	0.50
4, 8*	20	16	0.5	—26300.	—697.	0.046		38.6	37.8	0.028	700.
1, 7	20	16	1.0	— 14.0	— 0.664	0.075	2.164	0.99	0.940	0.062	14.15
2, 6	20	16	1.0	— 27.9	— 3.22	0.19	0.0955	2.58	2.23	0.26	20.02
3, 5*	20	16	1.0	0.012	— 0.0037	0.40		0.135	0.105	†	0.50
4, 8*	20	16	1.0	—20200.	—953.	0.075		37.6	35.7	0.062	539.
1, 7	20	16	2.0	— 4.3	— 0.715	0.14	1.829	1.00	0.877	0.137	8.95
2, 6	20	16	2.0	— 21.7	— 2.958	0.24	0.0775	2.24	1.84	0.29	20.02
3, 5*	20	16	2.0	0.0095	— 0.0038	0.52		0.115	0.082	†	0.50
4, 8*	20	16	2.0	— 5600.	—917.	0.14		36.0	31.6	0.137	323.

*For Case (5), stiffnesses and forces are four times, currents and flux densities are twice those tabulated. For Case (8), stiffnesses and forces are one-quarter, currents and flux densities are half those tabulated.

† Too flat to locate, but beyond x_{nF}.

‡ Exception: $x_{nF} > x_{nB}$, but $\mathscr{B}_{n2F} \approx \mathscr{B}_{n2m}$.

For Case (1) or (7) the maximum coil-2 voltage occurs at displacement

$$x_n = \frac{(Q_L - Q)Q - 1}{(Q_L - Q)(Q_0 - Q) + 1}.$$

Hence maximum coil-2 voltage can be obtained by computing ωL_2 for the appropriate x_n and multiplying the coil impedance Z_{L2} by the corresponding I_2. For Case (1) the maximum capacitor voltage occurs for maximum I_2 and can be computed by dividing I_2 by ωC. For Case (7), or for any of the bridge-circuit connections, the spotting of the maximum capacitor voltages is difficult. The voltage across the capacitor in the bridge-detector position is proportional to $I_2 - I_1$, the maximum of which must be less than maximum I_2, and the voltage across the capacitor in series with the source is proportional to $I_2 + I_1$, the maximum of which must be less than maximum $2I_2$. Since the capacitive reactance in each instance is $1/2\omega C$, the maximum capacitor voltages for Case (7) must be less than the maximum capacitor voltage for Case (1), especially for the capacitor in the bridge-detector position, which is an advantage of the bridge-circuit connection.

For Case (4) or (8) the maximum coil-2 voltage occurs at

$$x_n = \frac{Q}{Q_0 - Q},$$

and for Case (4) the maximum capacitor voltage is the same as the maximum voltage across a coil pair. The maximum coil-2 voltage can be obtained by computing ωL_2 for the appropriate x_n and multiplying the coil impedance Z_{L2} by the corresponding I_2. For Case (8) the voltage across the capacitor in the bridge-detector position is equal to the difference of the voltages across adjacent arms, $V_2 - V_1$, the maximum of which must be less than maximum V_2. But the voltage across the capacitor in parallel with the source is proportional to $V_2 + V_1$, the maximum of which can be substantially larger than maximum V_2 but not twice as large. Further, for equal current sources, V_2 for the bridge circuit, Case (8), is half V_2 for the circuit of Case (4). Hence the maximum capacitor voltages for Case (8) must be less than the maximum capacitor voltage for Case (4), especially for the capacitor in the bridge-detector position.

For Case (2) or (6) the analytic solution for the maximum coil-2 voltage becomes quite involved algebraically. Compared with Case (1) or (7), the curves of current versus displacement for Case (2) or (6) are less steep and have not such sharp peaks. Since the coil impedance Z_{L2} decreases with displacement, the maximum coil-2 voltage $I_2 Z_{L2}$ occurs at a smaller displacement than the displacement for maximum I_2, but use of Z_{L2} corresponding to maximum I_2 makes only a few percent error in the maximum coil-2 voltage V_2. For Case (2) the maximum capacitor voltage corresponds to the maximum coil-2 current, and for Case (6) the maximum capacitor voltage, being proportional to $I_2 - I_1$, and the reactance being $1/2\omega C$, must be substantially smaller than the capacitor voltage for Case (2).

For Case (3) or (5) the analytical solution for maximum coil-2 voltage also becomes quite involved algebraically. The curve of current I_{L2} with respect to displacement is very flat, so that multiplication of maximum I_{L2} by the corresponding Z_{L2} gives a large error. In fact, in some instances mathematical maximum I_{L2} occurs for $x_n > 1$, in a fictitious region. However, for Case (3) the quiescent coil voltage is $V/4$, and as the length of gap 2 increases the maximum coil-2 voltage reached must be less than $V/2$; for Case (5) the quiescent coil voltage is $V/2$, and as the length of gap 2 increases the maximum coil-2 voltage reached must be less than V. For Case (3) the capacitor voltage always is the same as the voltage across a coil pair, so that the maximum capacitor voltage is double the maximum coil-2 voltage but less than source voltage V. For Case (5) the voltage across the capacitor is equal to the difference of the voltages across adjacent arms, $V_2 - V_1$, the maximum of which must be less than maximum V_2. For these cases the coil and capacitor voltages are relatively low, less than the source voltage V or even less than $V/2$. However, to achieve force and stiffness comparable to the forces and stiffnesses that result from other connections, the source voltage for Case (3) or (5) must be relatively quite high but, nevertheless, serves as a good gauge of the maximum coil and capacitor voltages.

From a practical standpoint, some of the tabulated comparisons may be deceiving unless one bears in mind the bases used for normalization. For example, in comparing performance when a voltage-source excitation is used with performance when a current source is used, some relation between voltage- and current-source magnitudes must be assumed, which in this treatment is $V = RI$. For an established practical voltage source, the corresponding current source might be impractical, or vice versa. In the comparisons of Case (1) or (7) with Case (3) or (5) or the comparisons of Case (2) or (6) with Case (4) or (8), the differences seem to be enormous.

For Case (1), voltage source with series tuning, the coil currents at zero displacement are

$$I_0 = \frac{V}{2R} \frac{1}{\sqrt{Q^2 + 1}}$$

and increase or decrease from that value with displacement; whereas for Case (3), voltage source with parallel tuning, the coil currents at zero displacement are

$$I_0 = \frac{V}{4R} \frac{1}{\sqrt{Q_L^2 + 1}}$$

independent of the tuning, as indicated by absence of Q, and increase or decrease from that value with displacement. For example, for $Q_L = 15$ and $Q = 1$, $I_0 = V/60.3R$ for Case (3), and $I_0 = V/2.82R$ for Case (1), about 21.4 times as much as the quiescent current for Case (3), for equal voltage sources. Hence, in practice, if the voltage source used for Case (3) brings the coil currents to the tolerable limit, the voltage source used for Case (1) would have to be much smaller.

For Case (2), current source with series tuning, the coil currents at zero displacement are simply

$$I_0 = \frac{I}{2}$$

regardless of the tuning and increase or decrease from that value with displacement; whereas for Case (4), current source with parallel tuning, the coil currents at zero displacement are

$$I_0 = \frac{I}{2} \frac{2(Q_L - Q)}{\sqrt{Q^2 + 1}}$$

and increase or decrease from that value with displacement. For parallel tuning the individual coil and capacitor currents can be many times the source current. For example, for $Q_L = 15$ and $Q = 1$, $I_0 = 9.88I$ for Case (4), about 19.8 times as much as the quiescent current for Case (2) for equal current sources. Hence, in practice, if the current source used for Case (2) brings the coil current to the tolerable limit, the current source used for Case (4) would have to be much smaller.

The forces for Case (3) are much smaller than the forces for Case (1), as indicated by comparison of Figures 3-11 and 2-4. To emphasize the comparison, the force curves of Figure 3-11 first are plotted to the same ordinate scale as the force curves of Figure 2-4 and then are plotted with the ordinate scale expanded by a factor of 10, so that the shapes of the curves can be seen. The stiffness at zero displacement is much greater for Case (1) or (7) than for Case (3) or (5), as indicated by the force curves or by comparison of Figures 2-13 and 3-14, but the peaks of the force curves for Cases (3) and (5) occur at much larger displacements than the displacements for the peaks of the force curves for Cases (1) and (7). Even if the applied voltage for Cases (3) and (5) were increased to raise the force peaks to match the peaks for Cases (1) and (7), the displacements at which the peaks occur would not be affected. The coil currents and consequently the flux densities and forces for Cases (3) and (5) are quite different as functions of displacement from the corresponding functions for Cases (1) and (7), as examination of the equations and the corresponding plots indicates.

The forces for Case (4) are much larger than the forces for Case (2), as indicated by comparison of Figures 3-3 and 3-16. Hence the curves of Figure 3-16 cannot be plotted to the same ordinate scale as adopted for the curves of Figure 3-3, for direct comparison. The stiffness at zero displacement is much greater for Case (4) or (8) than for Case (2) or (6), as indicated by the force curves or by comparison of Figures 3-9 and 3-22, but the peaks of the force curves for Cases (2) and (6) occur at considerably larger displacements than the displacements for the peaks of the force curves for Cases (4) and (8). The locations of these peaks are independent of the magnitudes of the current sources. The coil currents and consequently the flux densities and forces for Cases (4) and (8) are quite different as functions of displacement from the corresponding functions for Cases (2) and (6), as examination of the equations and the corresponding plots indicates.

Though the bridge-circuit connections, Cases (5) through (8), theoretically give the same coil current, flux density, and force relations as result from the connections for Cases (1) through (4), as tabulated at the end of Chapter 4, the bridge-circuit connection has the considerable advantage that drift in capacitance has no effect on the quiescent position of the suspended member.

Obviously, if a suspension connected as for Case (1) or (7) gives satisfactory operation, the same suspension connected as for Case (3) or (5) would be most unlikely to give satisfactory operation with the same voltage source. Likewise, if a suspension connected as for Case (4) or (8) gives satisfactory operation, the same suspension connected as for Case (2) or (6) would be most unlikely to give satisfactory operation with the same current source. Finally, exchange of voltage and current source through the relation $V = RI$ may or may not be practical. Use of Table 5-1 permits study of the ratios by which the voltages or currents, applied and appearing in the circuits, and the fluxes and flux densities must change if the restoring forces or the stiffnesses of two cases are to be adjusted to different levels by change in applied voltage or current, since forces and stiffnesses vary as V^2 or I^2, and the other voltages and currents and the fluxes and flux densities vary as V or I. The changes theoretically required may not always be within the heating limit of the conductors, the breakdown limit of insulation, or the saturation limit of magnetic materials. Other bases of comparison of mode of operation may be preferable to constant V, constant I, and $V = RI$.

5-2 Fixed Limit on Gap Flux Density

A practical limit on the operation of the magnetic suspension as analyzed thus far is the effect of saturation of magnetic material. Presumably, without flux bottlenecks elsewhere in the magnetic structure and with due attention to the distribution of leakage fluxes, the maximum flux density occurs in the longer gap, or gap 2 of Figure 2-1, in accordance with the conventions adopted for the analysis. With very small gap length compared with cross-sectional dimensions, the flux density in the gap is essentially the same as the flux density at the magnetic surfaces adjacent to the gap, which therefore limits the flux density achievable in the gap. As magnetic saturation is approached at these surfaces, the flux density in the gap cannot continue in proportion to the coil-2 current, and the solution as made becomes invalid.

However, until the magnetic material requires an appreciable portion of the magnetomotive force in comparison with the magnetomotive force required for the gap, even though the magnetic material may be worked around the knee of its magnetization curve, the effect on the solution of the magnetic circuit may be negligible. Hence the practical limit on gap flux density analytically is the point at which the magnetomotive force for the magnetic material is judged to cease to be negligible. In this study, a normalized limit has arbitrarily been taken as $\mathcal{B}_{n2L} = 0.735$, which happened to be the maximum encountered in some of the early analyses. Actually this number serves merely as a basis for the comparisons in Table 5-2 and for the curves of Figures 5-1 through 5-23. For a particular suspension, some limit depending on the magnetic material involved would be placed on $\mathcal{B}_2$, which in turn would determine the limit on $\mathcal{B}_{n2}$, and that limit would be used instead of 0.735 as arbitrarily selected. The result would be to multiply all voltages and

Table 5-2. Equal Flux-Density Limits $\mathscr{B}_{n2m} = \mathscr{B}_{n2F} = 0.735$

Case	Q_L	Q_0	Q	$\dot{F}_{n0}$	F_{nm}	V_{nRL0}	I_{nLF}	V or I
1, 7	10	6	0.5	−2.75	−0.189	7.10	0.780	0.789
2, 6	10	6	0.5	−1.44	−0.220	4.38	0.973	0.436
3, 5	10	6	0.5	−0.828	−0.179	3.30	1.05	6.60
4, 8	10	6	0.5	−2.75	−0.189	7.10	0.780	0.0415
1, 7	10	6	1.0	−2.99	−0.356	6.16	0.864	0.864
2, 6	10	6	1.0	−1.90	−0.379	4.81	0.981	0.478
3, 5	10	6	1.0	−1.41	−0.337	3.77	1.130	7.54
4, 8	10	6	1.0	−2.99	−0.356	6.16	0.864	0.0481
1, 7	10	6	2.0	−1.29	−0.454	4.81	1.063	1.070
2, 6	10	6	2.0	−1.75	−0.463	5.65	1.062	0.559
3, 5	10	6	2.0	−1.34	−0.480	4.48	1.210	8.95
4, 8	10	6	2.0	−1.29	−0.454	4.81	1.063	0.0669
1, 7	15	10	0.5	−5.65	−0.247	10.36	0.762	0.770
2, 6	15	10	0.5	−1.91	−0.234	5.40	0.915	0.359
3, 5	15	10	0.5	−0.785	−0.170	3.68	1.030	7.35
4, 8	15	10	0.5	−5.65	−0.247	10.36	0.762	0.0265
1, 7	15	10	1.0	−5.28	−0.398	8.64	0.812	0.812
2, 6	15	10	1.0	−2.46	−0.390	6.05	0.925	0.402
3, 5	15	10	1.0	−1.25	−0.298	4.18	1.003	8.36
4, 8	15	10	1.0	−5.28	−0.398	8.64	0.812	0.0290
1, 7	15	10	2.0	−2.00	−0.485	6.15	0.912	0.912
2, 6	15	10	2.0	−2.79	−0.489	7.11	0.969	0.472
3, 5	15	10	2.0	−1.31	−0.426	5.03	1.090	10.06
4, 8	15	10	2.0	−2.00	−0.485	6.15	0.912	0.0351
1, 7	20	16	0.5	−9.91	−0.264	13.54	0.750	0.758
2, 6	20	16	0.5	−2.43	−0.228	6.05	0.842	0.302
3, 5	20	16	0.5	−0.435	−0.126	3.30	1.01	6.60
4, 8	20	16	0.5	−9.91	−0.264	13.54	0.750	0.0194
1, 7	20	16	1.0	−8.60	−0.405	11.08	0.775	0.782
2, 6	20	16	1.0	−3.23	−0.373	6.82	0.878	0.340
3, 5	20	16	1.0	−0.708	−0.222	3.85	1.030	7.69
4, 8	20	16	1.0	−8.60	−0.405	11.08	0.775	0.0206
1, 7	20	16	2.0	−3.03	−0.500	7.51	0.839	0.839
2, 6	20	16	2.0	−3.50	−0.477	8.05	0.900	0.402
3, 5	20	16	2.0	−0.855	−0.343	4.75	1.070	9.50
4, 8	20	16	2.0	−3.03	−0.500	7.51	0.839	0.0233

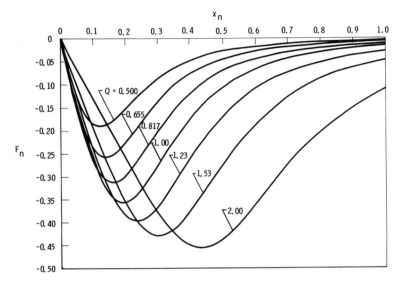

Figure 5-1 Normalized centering force versus normalized displacement, with limited maximum flux density, Case (1), (4), (7), or (8); $Q_0 = 6$.

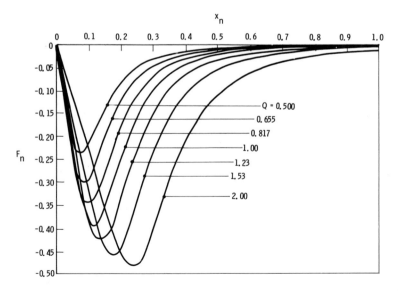

Figure 5-2 Normalized centering force versus normalized displacement, with limited maximum flux density, Case (1), (4), (7), or (8); $Q_0 = 10$.

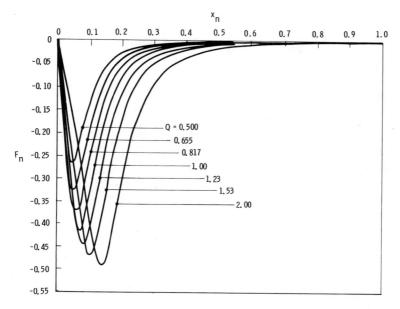

Figure 5-3 Normalized centering force versus normalized displacement, with limited maximum flux density, Case (1), (4), (7), or (8); $Q_0 = 16$.

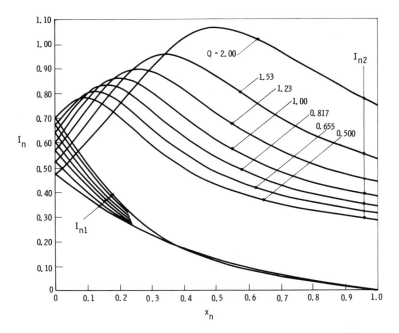

Figure 5-4 Normalized coil currents versus normalized displacement, with limited maximum flux density, Case (1), (4), (7), or (8); $Q_0 = 6$.

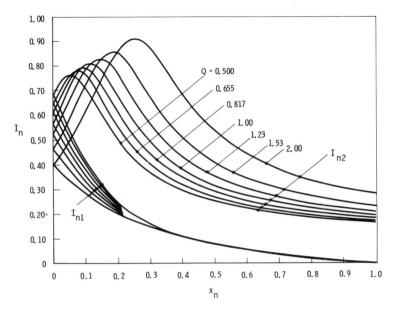

Figure 5-5 Normalized coil currents versus normalized displacement, with limited maximum flux density, Case (1), (4), (7), or (8); $Q_0 = 10$.

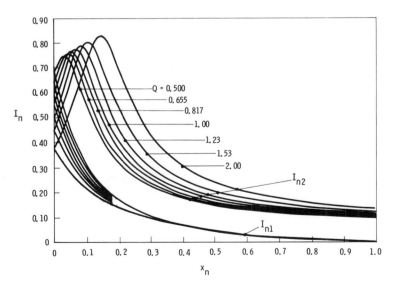

Figure 5-6 Normalized coil currents versus normalized displacement, with limited maximum flux density, Case (1), (4), (7), or (8); $Q_0 = 16$.

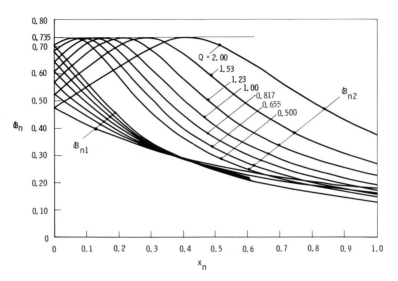

Figure 5-7 Normalized flux densities versus normalized displacement, with limited maximum flux density, Case (1), (4), (7), or (8); $Q_0 = 6$.

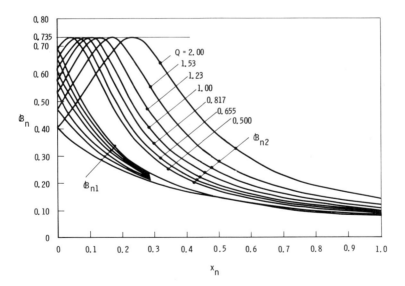

Figure 5-8 Normalized flux densities versus normalized displacement, with limited maximum flux density, Case (1), (4), (7), or (8); $Q_0 = 10$.

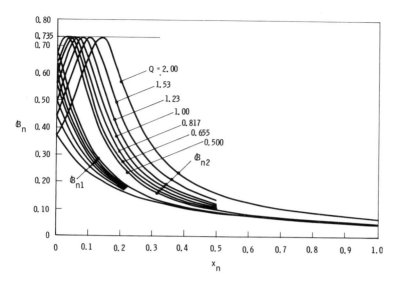

Figure 5-9 Normalized flux densities versus normalized displacement, with limited maximum flux density, Case (1), (4), (7), or (8); $Q_0 = 16$.

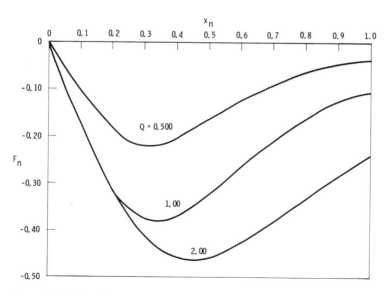

Figure 5-10 Normalized centering force versus normalized displacement, with flux density limited at maximum force, Case (2) or (6); $Q_0 = 6$.

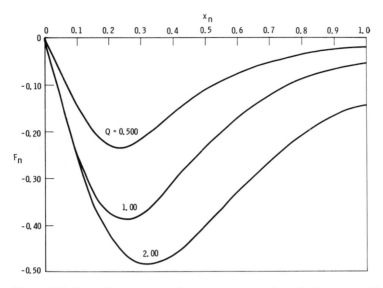

Figure 5-11 Normalized centering force versus normalized displacement, with flux density limited at maximum force, Case (2) or (6); $Q_0 = 10$.

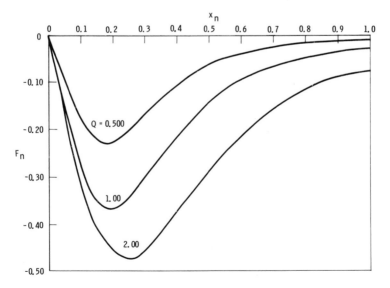

Figure 5-12 Normalized centering force versus normalized displacement, with flux density limited at maximum force, Case (2) or (6); $Q_0 = 16$.

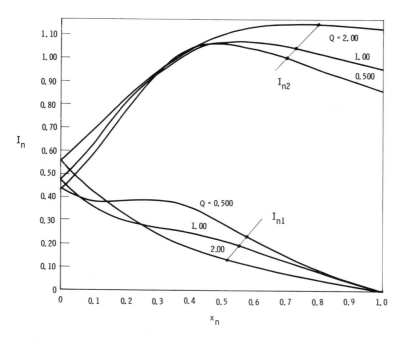

Figure 5-13 Normalized coil currents versus normalized displacement, with flux density limited at maximum force, Case (2) or (6); $Q_0 = 6$.

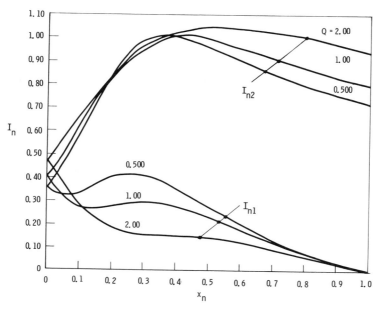

Figure 5-14 Normalized coil currents versus normalized displacement, with flux density limited at maximum force, Case (2) or (6); $Q_0 = 10$.

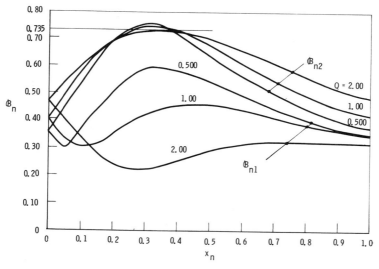

Figure 5-15 Normalized flux densities versus normalized displacement, with flux density limited at maximum force, Case (2) or (6); $Q_0 = 10$.

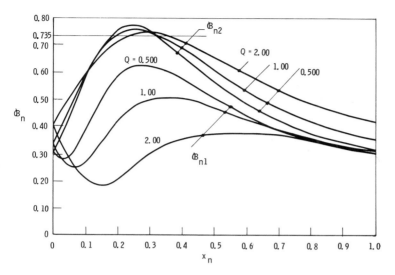

Figure 5-16 Normalized flux densities versus normalized displacement, with flux density limited at maximum force, Case (2) or (6); $Q_0 = 16$.

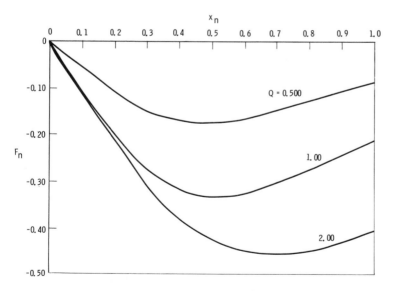

Figure 5-17 Normalized centering force versus normalized displacement, with flux density limited at maximum force, Case (3) or (5); $Q_L = 10$, $Q_0 = 6$.

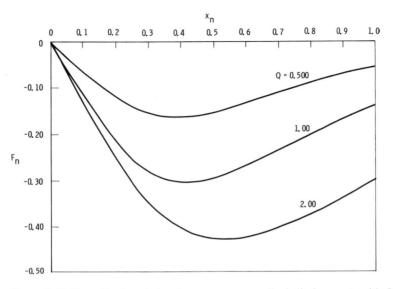

Figure 5-18 Normalized centering force versus normalized displacement, with flux density limited at maximum force, Case (3) or (5); $Q_L = 15$, $Q_0 = 10$.

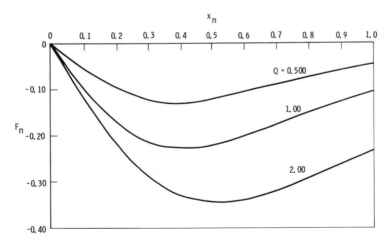

Figure 5-19 Normalized centering force versus normalized displacement, with flux density limited at maximum force, Case (3) or (5); $Q_L = 20$, $Q_0 = 16$.

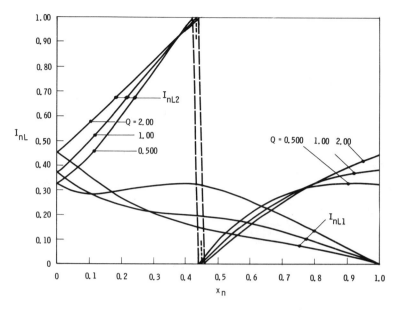

Figure 5-20 Normalized coil currents versus normalized displacement, with flux density limited at maximum force, Case (3) or (5); $Q_L = 10$, $Q_0 = 6$.

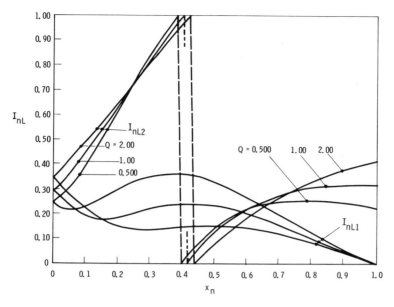

Figure 5-21 Normalized coil currents versus normalized displacement, with flux density limited at maximum force, Case (3) or (5); $Q_L = 15$, $Q_0 = 10$.

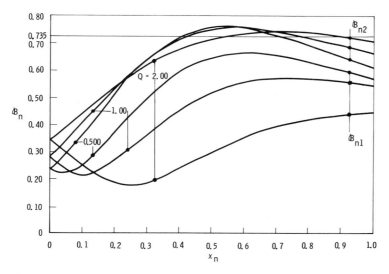

Figure 5-22 Normalized flux densities versus normalized displacement, with flux density limited at maximum force, Case (3) or (5); $Q_L = 15$, $Q_0 = 10$.

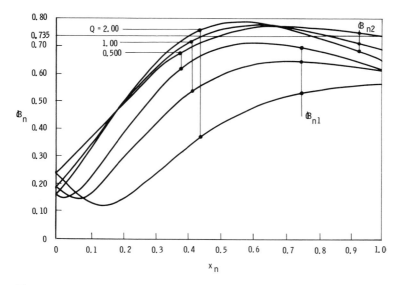

Figure 5-23 Normalized flux densities versus normalized displacement, with flux density limited at maximum force, Case (3) or (5); $Q_L = 20$, $Q_0 = 16$.

currents in Table 5-2 and the ordinates of current and flux-density curves by $\mathscr{B}_{n2L}/0.735$, and to multiply all forces and stiffnesses by $(\mathscr{B}_{n2L}/0.735)^2$.

For Cases (1) or (7) and (4) or (8) maximum flux density $\mathscr{B}_2$ occurs at somewhat smaller displacement x than the displacement for maximum restoring force F, whereas for Cases (2) or (6) and (3) or (5) the situation is almost always reversed. In translating the curves from equal applied voltage or current bases to equal flux-density limit bases, the limit 0.735 is applied to maximum $\mathscr{B}_{n2}$ if it occurs at a displacement x_{nB} smaller than the displacement corresponding to maximum restoring force F, but if maximum $\mathscr{B}_{n2}$ occurs at a displacement x_{nB} larger than the displacement x_{nF} corresponding to maximum restoring force F, then the limit 0.735 is applied to the flux density $\mathscr{B}_{n2}$ that corresponds to maximum F. These flux densities are designated $\mathscr{B}_{n2m}$ and $\mathscr{B}_{n2F}$, respectively. The fact that $\mathscr{B}_{n2F}$ is exceeded somewhat as displacement is increased from the point corresponding to maximum F is not a particular concern, because beyond that point such modification in performance that may be caused by further approach to magnetic saturation can extend only over a small displacement region of interest primarily for erection and is relatively minor. Hence in obtaining the curves of Figures 5-1 through 5-23 from the corresponding curves for operation with constant voltage or current sources and the numbers in Table 5-2, the ratio $0.735/\mathscr{B}_{n2m}$ or $0.735/\mathscr{B}_{n2F}$ is used for currents, voltages, or flux densities, and their squares are used for forces or stiffnesses.

The curves of stiffness at zero displacement, $\dot{F}_{n0}$ versus Q, Figures 2-13, 3-9, 3-14, and 3-22, are not translated to the fixed $\mathscr{B}_{n2} = 0.735$ level. However, selected translated stiffness data are shown in Table 5-2, and factors for translating any of the data from a fixed source V or I basis to a fixed $\mathscr{B}_{n2} = 0.735$ basis are given in Table 5-3. When translated to the same flux-density level, the respective curves for Cases (3) and (5) become identical, and those for Cases (4) and (8) become identical. The translation factors for Case (5) are half the translation factors for Case (3), and the translation factors for Case (8) are twice the translation factors for Case (4).

The translation of the curves for all cases to substantially the same maximum air-gap flux-density level permits plotting them to the same respective scales of ordinates and thus greatly facilitates intercomparisons. In particular, double scales for Cases (3) and (5) and for Cases (4) and (8) are not needed, and these cases give forces, currents, and flux densities of the same orders of magnitude as Cases (1) or (7) and (2) or (6). Not only do Cases (3) and (5) merge as well as Cases (4) and (8), but Cases (4) and (8) merge with Cases (1) and (7),

Table 5-3. Translation Factors

Q	Cases (1) and (7): $0.735/\mathcal{B}_{n2m}$			Cases (2) and (6): $0.735/\mathcal{B}_{n2F}$		
	$Q_0 = 6$	10	16	$Q_0 = 6$	10	16
0.500	0.789	0.770	0.758	0.436	0.359	0.302
0.655	0.811	0.782	0.765	0.448	0.371	0.315
0.817	0.835	0.795	0.773	0.462	0.388	0.324
1.000	0.864	0.812	0.782	0.478	0.402	0.340
1.230	0.905	0.832	0.795	0.498	0.420	0.355
1.530	0.963	0.862	0.810	0.523	0.441	0.373
2.000	1.070	0.912	0.839	0.559	0.472	0.402

Q	Case (3): $0.735/\mathcal{B}_{n2F}$			Case (4): $0.735/\mathcal{B}_{n2m}$		
	$Q_L = 10$ $Q_0 = 6$	15 10	20 16	$Q_L = 10$ $Q_0 = 6$	15 10	20 16
0.500	6.60	7.35	6.60	0.0415	0.0265	0.0194
0.655	6.87	7.57	6.87	0.0434	0.0273	0.0198
0.817	7.17	7.90	7.17	0.0455	0.0281	0.0202
1.000	7.54	8.36	7.69	0.0481	0.0290	0.0206
1.230	7.90	8.95	8.07	0.0516	0.0303	0.0212
1.530	8.30	9.41	8.75	0.0568	0.0320	0.0220
2.000	8.95	10.06	9.50	0.0669	0.0351	0.0233

The translation factors for Case (5) are half as large as for Case (3).
The translation factors for Case (8) are twice as large as for Case (4).

so that when concerned with force, stiffness, coil currents, and voltages, only three of the eight different circuits examined give different results. For Case (4), the condition for imposing the same maximum flux-density limit as for Case (1) is

$$\frac{V}{R} = IQ_c = I(Q_L - Q),$$

which reduces Equation (3-25) to its original form, Equation (3-23), the normalized form of which is identical with Equation (2-7). The source voltages and currents in Table 5-2 show how the source voltages or currents must change, based on whatever fixed source voltages and currents are held for Cases (1) through (8), with results shown in Table 5-1, to achieve operation with fixed flux-density limit.

In Table 5-2, and in following tables, x_{nF} and x_{nB} are not repeated, because for a fixed set of Q_L, Q_0, and Q they do not change. For Cases (1) and (4)

$$x_{nB} = \frac{(Q_0 - Q)Q - 1}{(Q_0 - Q)^2 + 1},$$

but for Cases (2) and (3) the algebraic expressions are quite complicated. For Case (2) x_{nB} is located from the plotted $\mathscr{B}_{n2}$ curves, but for Case (3) the curves are too flat to locate the maxima with any certainty. For Cases (2) and (3) x_{nF} is located from the plotted F_n curves, and the corresponding $\mathscr{B}_{n2F}$ is located from the plotted $\mathscr{B}_{n2}$ curves.

5-3 Other Bases of Comparison

In Tables 5-4 through 5-6, comparative operating data are shown on the bases of equal coil currents for maximum restoring forces, equal stiffnesses at zero displacement, and equal maximum forces. Of course, plots could be made on these bases that would be instructive, but the number of them would multiply beyond practical bounds. In these tables arbitrary limits of $I_{nLF} = 1.00$, $F_{n0} = 10.0$, and $F_{nm} = 1.00$ are chosen. For equal flux-density limits, Table 5-2, $\mathscr{B}_{n2m} = \mathscr{B}_{n2F} = 1.00$ might more logically have been chosen, but many computer curves using the limit 0.735 had been run before a wider look at the situation had been taken.

Another very interesting basis of comparison comes from the use of the pull on one side of the block, Figure 2-1, under quiescent conditions (block centered) as the force base. The force equations then translate to

$$F_{Vs} = \frac{N^2 V^2 \mu A}{4 R^2 g_0^2 (Q^2 + 1)} \; \frac{-4\left(\dfrac{Q_0 Q}{Q^2 + 1} - 1\right) x_n}{1 - 2\left[\dfrac{Q_0^2(Q^2 - 1)}{(Q^2 + 1)^2} - \dfrac{2 Q_0 Q}{Q^2 + 1} + 1\right] x_n^2 + \left[\dfrac{Q_0(Q_0 - 2Q)}{Q^2 + 1} + 1\right]^2 x_n^4} \tag{5-1}$$

for Case (1), to

$$F_{Is} = \frac{N^2 I^2 \mu A}{4 g_0^2} \; \frac{-4\left(\dfrac{Q_0 Q}{Q^2 + 1} - 1\right) x_n}{1 + 2\left(\dfrac{Q_0 Q}{Q^2 + 1} - 1\right) x_n^2 + \left[\dfrac{Q_0(Q_0 - 2Q)}{Q^2 + 1} + 1\right] x_n^4} \tag{5-2}$$

for Case (2), to

$$
F_{Vp} = \frac{N^2 V^2 \mu A}{16 R^2 g_0^2 (Q_L^2 + 1)} \frac{-4\left(\dfrac{Q_0 Q}{Q^2+1} - 1\right) x_n}{1 - 2\left\{1 - \dfrac{Q_0[Q_L + Q_0 + Q + Q_L Q(Q_L - Q_0 + Q)]}{(Q_L^2+1)(Q^2+1)}\right\} x_n^2 + \left\{1 + \dfrac{\begin{array}{c} Q_0(Q_L - Q_0 + Q) \\ \times [Q_0(Q_L - Q_0 + Q) - 2(Q_L Q + 1)] \end{array}}{(Q_L^2+1)(Q^2+1)}\right\} x_n^4}
$$

$$(5\text{-}3)$$

for Case (3), and to

$$
F_{Ip} = \frac{N^2 I^2 \mu A}{g_0^2} \frac{(Q_L - Q)^2}{Q^2 + 1}
$$

$$
\times \frac{-4\left(\dfrac{Q_0 Q}{Q^2+1} - 1\right) x_n}{1 - 2\left[\dfrac{Q^2(Q^2-1)}{(Q^2+1)^2} - \dfrac{2Q_0 Q}{Q^2+1} + 1\right] x_n^2 + \left[\dfrac{Q_0(Q_0 - 2Q)}{Q^2+1} + 1\right]^2 x_n^4}
$$

$$(5\text{-}4)$$

for Case (4). These transpositions accomplish two things: (1) the normalized stiffnesses at zero displacement are identical, and (2) each of the denominators of the normalized forces has the form $1 + $ a function of x_n, which directly indicates the nonlinearity of the expression. Further, Equation (5-2) for Case (2) actually has undergone no change with respect to Equation (3-2) except the division of numerator and denominator by $Q^2 + 1$; and $N^2 I^2 \mu A / 4 g_0^2$, the quiescent pull on one side of the block, is independent of Q_L or Q. Equation (3-8),

$$
\dot{F}_{n0} = \frac{4(Q^2 - Q_0 Q + 1)}{Q^2 + 1} = -4\left(\frac{Q_0}{Q + \dfrac{1}{Q}} - 1\right),
$$

therefore, is the normalized stiffness equation for all the translated cases, and its plot, Figure 3-9, represents all the translated cases. By plotting on a logarithmic scale, Figure 5-24, the curves are made symmetrical about $Q = 1$.

Table 5-4. Equal Coil Currents at Maximum Force, $I_{nLF} = 1.00$

Case	Q_L	Q_0	Q	$\dot{F}_{n0}$	F_{nm}	V_{nRL0}	$\mathcal{B}_{n2m}$ or $\mathcal{B}_{n2F}$*	V or I
1, 7	10	6	0.5	− 4.5	−0.310	9.08	0.940	1.01
2, 6	10	6	0.5	− 1.53	−0.233	4.52	0.755	0.429
3, 5	10	6	0.5	− 0.74	−0.160	3.13	0.700	6.25
4, 8	10	6	0.5	− 4.5	−0.310	9.08	0.940	0.0531
1, 7	10	6	1.0	− 4.0	−0.474	7.10	0.850	1.00
2, 6	10	6	1.0	− 1.97	−0.394	4.90	0.749	0.488
3, 5	10	6	1.0	− 1.11	−0.264	3.33	0.650	6.67
4, 8	10	6	1.0	− 4.0	−0.474	7.10	0.850	0.0555
1, 7	10	6	2.0	− 1.14	−0.405	4.54	0.695	1.01
2, 6	10	6	2.0	− 1.55	−0.409	5.28	0.690	0.525
3, 5	10	6	2.0	− 0.92	−0.313	3.70	0.607	7.40
4, 8	10	6	2.0	− 1.14	−0.405	4.54	0.695	0.0632
1, 7	15	10	0.5	− 9.8	−0.425	13.60	0.962	1.01
2, 6	15	10	0.5	− 2.28	−0.278	5.91	0.805	0.392
3, 5	15	10	0.5	− 0.74	−0.160	3.57	0.715	7.15
4, 8	15	10	0.5	− 9.80	−0.425	13.60	0.962	0.0348
1, 7	15	10	1.0	− 8.0	−0.605	10.64	0.907	1.00
2, 6	15	10	1.0	− 2.88	−0.456	6.55	0.796	0.435
3, 5	15	10	1.0	− 1.23	−0.295	4.17	0.733	8.33
4, 8	15	10	1.0	− 8.0	−0.605	10.64	0.907	0.0357
1, 7	15	10	2.0	− 2.4	−0.583	6.73	0.806	1.01
2, 6	15	10	2.0	− 2.97	−0.521	7.33	0.758	0.488
3, 5	15	10	2.0	− 1.11	−0.361	4.63	0.677	9.27
4, 8	15	10	2.0	− 2.4	−0.583	6.73	0.806	0.0384
1, 7	20	16	0.5	−17.6	−0.469	18.11	0.980	1.01
2, 6	20	16	0.5	− 3.43	−0.321	7.19	0.872	0.358
3, 5	20	16	0.5	− 0.43	−0.124	3.27	0.732	6.53
4, 8	20	16	0.5	−17.6	−0.469	18.11	0.980	0.0259
1, 7	20	16	1.0	−14.3	−0.675	14.30	0.948	1.01
2, 6	20	16	1.0	− 4.20	−0.483	7.76	0.840	0.388
3, 5	20	16	1.0	− 0.66	−0.206	3.70	0.707	7.40
4, 8	20	16	1.0	−14.3	−0.675	14.30	0.948	0.0266
1, 7	20	16	2.0	− 4.3	−0.713	8.97	0.878	1.00
2, 6	20	16	2.0	− 4.32	−0.588	8.93	0.815	0.447
3, 5	20	16	2.0	− 0.72	−0.287	4.35	0.674	8.70
4, 8	20	16	2.0	− 4.3	−0.713	8.97	0.878	0.0278

*For Cases (2) or (6) and (3) or (5).

Table 5-5. Equal Maximum Stiffness $\dot{F}_{n0} = 10.0$

Case	Q_L	Q_0	Q	F_{nm}	V_{nRL0}	I_{nLF0}	$\mathscr{B}_{n2m}$ or $\mathscr{B}_{n2F}$*	V or I
1, 7	10	6	0.5	−0.688	13.57	1.49	1.41	1.51
2, 6	10	6	0.5	−1.53	15.55	2.56	1.94	1.15
3, 5	10	6	0.5	−2.16	11.43	3.58	2.57	22.96
4, 8	10	6	0.5	−0.688	13.57	1.49	1.41	0.0792
1, 7	10	6	1.0	−1.19	11.24	1.58	1.35	1.58
2, 6	10	6	1.0	−2.00	11.05	2.25	1.69	1.10
3, 5	10	6	1.0	−2.39	10.04	3.01	1.96	20.08
4, 8	10	6	1.0	−1.19	11.24	1.58	1.35	0.0878
1, 7	10	6	2.0	−3.40	13.47	2.97	2.06	2.99
2, 6	10	6	2.0	−2.66	13.45	2.54	1.77	1.35
3, 5	10	6	2.0	−3.43	12.23	3.30	2.00	24.45
4, 8	10	6	2.0	−3.40	13.47	2.97	2.06	0.0187
1, 7	15	10	0.5	−0.432	13.71	1.02	0.98	1.03
2, 6	15	10	0.5	−1.22	12.40	2.10	1.69	0.822
3, 5	15	10	0.5	−2.16	13.13	3.68	2.63	26.26
4, 8	15	10	0.5	−0.432	13.71	1.02	0.98	0.0352
1, 7	15	10	1.0	−0.753	11.89	1.12	1.02	1.12
2, 6	15	10	1.0	−1.59	12.23	1.87	1.49	0.812
3, 5	15	10	1.0	−2.38	11.87	2.85	2.09	23.73
4, 8	15	10	1.0	−0.753	11.89	1.12	1.02	0.0399
1, 7	15	10	2.0	−2.44	13.74	2.04	1.65	2.04
2, 6	15	10	2.0	−1.75	13.50	1.84	1.40	0.895
3, 5	15	10	2.0	−3.26	13.93	3.01	2.04	27.86
4, 8	15	10	2.0	−2.44	13.74	2.04	1.65	0.0785
1, 7	20	16	0.5	−0.266	13.66	0.753	0.738	0.760
2, 6	20	16	0.5	−0.935	11.84	1.65	1.44	0.590
3, 5	20	16	0.5	−2.90	15.84	4.85	3.55	31.67
4, 8	20	16	0.5	−0.266	13.66	0.753	0.738	0.0195
1, 7	20	16	1.0	−0.473	12.00	0.838	0.796	0.847
2, 6	20	16	1.0	−1.153	12.05	1.55	1.30	0.600
3, 5	20	16	1.0	−3.14	14.45	3.90	2.76	28.90
4, 8	20	16	1.0	−0.473	12.00	0.838	0.796	0.0223
1, 7	20	16	2.0	−1.63	13.67	1.528	1.34	1.528
2, 6	20	16	2.0	−1.43	13.54	1.52	1.24	0.679
3, 5	20	16	2.0	−4.00	16.25	3.74	2.52	32.50
4, 8	20	16	2.0	−1.63	13.67	1.528	1.34	0.0422

*For Cases (2) or (6) and (3) or (5).

Table 5-6. Equal Maximum Forces $F_{nm} = 1.00$

Case	Q_L	Q_0	Q	$\dot{F}_{n0}$	V_{nRL0}	I_{nLF}	$\mathcal{B}_{n2m}$ or $\mathcal{B}_{n2F}$*	V or I
1, 7	10	6	0.5	-14.6	16.32	1.80	1.690	1.81
2, 6	10	6	0.5	$-\ 6.55$	20.82	2.07	1.566	0.929
3, 5	10	6	0.5	$-\ 4.63$	7.80	2.49	1.748	15.60
4, 8	10	6	0.5	-14.6	16.32	1.80	1.690	0.0958
1, 7	10	6	1.0	$-\ 8.43$	10.33	1.46	1.236	1.45
2, 6	10	6	1.0	$-\ 5.00$	7.81	1.59	1.194	0.777
3, 5	10	6	1.0	$-\ 4.17$	6.49	1.95	1.264	12.97
4, 8	10	6	1.0	$-\ 8.43$	10.33	1.46	1.236	0.0810
1, 7	10	6	2.0	$-\ 2.83$	7.14	1.57	1.090	1.59
2, 6	10	6	2.0	$-\ 3.78$	8.28	1.56	1.082	0.822
3, 5	10	6	2.0	$-\ 2.92$	6.63	1.79	1.087	13.25
4, 8	10	6	2.0	$-\ 2.83$	7.14	1.57	1.090	0.0991
1, 7	15	10	0.5	-23.2	20.95	1.55	1.480	1.56
2, 6	15	10	0.5	$-\ 8.16$	11.19	1.90	1.520	0.743
3, 5	15	10	0.5	$-\ 4.61$	8.93	2.50	1.785	17.85
4, 8	15	10	0.5	-23.2	20.95	1.55	1.480	0.0535
1, 7	15	10	1.0	-13.3	13.72	1.29	1.170	1.29
2, 6	15	10	1.0	$-\ 6.31$	9.70	1.48	1.170	0.645
3, 5	15	10	1.0	$-\ 4.18$	7.67	1.84	1.350	15.34
4, 8	15	10	1.0	-13.3	13.72	1.29	1.170	0.0460
1, 7	15	10	2.0	$-\ 4.11$	8.82	1.31	1.056	1.31
2, 6	15	10	2.0	$-\ 5.71$	10.20	1.39	1.053	0.677
3, 5	15	10	2.0	$-\ 3.07$	7.72	1.66	1.120	15.43
4, 8	15	10	2.0	$-\ 4.11$	8.82	1.31	1.056	0.0504
1, 7	20	16	0.5	-37.6	26.55	1.47	1.436	1.48
2, 6	20	16	0.5	-11.69	12.68	1.77	1.540	0.633
3, 5	20	16	0.5	$-\ 3.45$	9.30	2.84	2.081	18.6
4, 8	20	16	0.5	-37.6	26.55	1.47	1.436	0.0379
1, 7	20	16	1.0	-21.1	17.45	1.22	1.157	1.23
2, 6	20	16	1.0	$-\ 8.67$	11.20	1.44	1.210	0.558
3, 5	20	16	1.0	$-\ 3.31$	8.32	2.24	1.689	16.63
4, 8	20	16	1.0	-21.1	17.45	1.22	1.157	0.0324
1, 7	20	16	2.0	$-\ 6.06$	10.62	1.19	1.039	1.184
2, 6	20	16	2.0	$-\ 7.34$	11.65	1.30	1.063	0.582
3, 5	20	16	2.0	$-\ 2.49$	8.11	1.87	1.258	16.21
4, 8	20	16	2.0	$-\ 6.06$	10.62	1.19	1.039	0.0331

*For Cases (2) or (6) and (3) or (5).

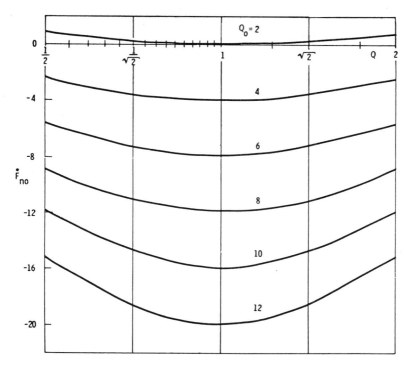

Figure 5-24 Universal logarithmic plot of normalized initial stiffness versus Q, for various values of Q_0, based on Equations (5-1) through (5-4).

If Equations (5-1) through (5-4) are written for $Q = 1$, which is a common operating condition, they are considerably simplified, and the nonlinearity of the denominators is more readily visualized. For Cases (1) and (4)

$$F_n = \frac{-2(Q_0 - 2)x_n}{1 + 2(Q_0 - 1)x_n^2 + \left[\dfrac{Q_0^2}{2} - (Q_0 - 1)\right]^2 x_n^4},$$

(5-5)

for Case (2)

$$F_n = \frac{-2(Q_0 - 2)x_n}{1 + (Q_0 - 1)x_n^2 + \left[\dfrac{Q_0^2}{2} - (Q_0 - 1)\right]x_n^4},$$

(5-6)

and for Case (3)

$$F_n = \cfrac{-2(Q_0 - 2)x_n}{1 + \left\{Q_0 - 2 + \cfrac{Q_0[(Q_L - 1)Q_0 - 2Q_L]}{Q_L^2 + 1}\right\}x_n^2 + \left\{1 + \cfrac{Q_0(Q_L - Q_0 + 1)[Q_0(Q_L - Q_0 + 1) - 2(Q_L + 1)]}{2(Q_L^2 + 1)}\right\}x_n^4} ,(5\text{-}7)$$

all normalized on the basis of the respective coefficients in Equations (5-1) through (5-4).

5-4 Nonsinusoidal Excitation

When magnetic suspensions are used in missiles, space vehicles, or in other applications in which space and energy requirements must be minimized, square voltage waves, which can be obtained very simply by chopping battery voltages, are commonly used because less space, apparatus, and auxiliary energy is required than for the generation of sinusoidal voltages. Passive magnetic suspensions in fact perform almost as well on square-wave excitation as with sinusoidal excitation.[40] Fourier analysis of a square wave shows that the harmonic voltages decrease very markedly with the order of the harmonic, and the corresponding coil current harmonics are further attenuated by the increased coil-circuit impedances for the various harmonics.

The impedance of one circuit of a series-tuned suspension at zero displacement, for angular frequency $h\omega_1$, is

$$Z_h = 2R\left(1 + j\frac{h\omega_1 L}{R} - j\frac{1}{2h\omega_1 RC}\right),$$

in which h is the order of the harmonic and ω_1 is the angular frequency of the fundamental of the source. If the circuit is tuned to the upper half-power point at fundamental frequency, then

$$\frac{\omega_1 L}{R} - \frac{1}{2\omega_1 RC} = 1;$$

and the impedance can be written

$$Z_h = 2R\left\{1 + jQ_{L1}\left[h - \frac{1}{h}\left(1 - \frac{1}{Q_{L1}}\right)\right]\right\},$$

in which

$$Q_{L1} = \omega_1 \frac{L}{R}.$$

If the source is a square voltage wave having a fundamental component of rms value V_1, the harmonics have rms values V_1/h, in which h is odd. The quiescent force components corresponding to the various harmonics are

$$F_h = \frac{N^2 V_1^2 \mu A}{g_0^2 h^2 Z_h^2} = \frac{N^2 V_1^2 \mu A}{4R^2 g_0^2} \frac{1}{h^2 \left\{ 1 + Q_{L1}^2 \left[h - \frac{1}{h}\left(1 - \frac{1}{Q_{L1}} \right) \right]^2 \right\}}$$

$$= \frac{2F_1}{h^2 \left\{ 1 + Q_{L1}^2 \left[h - \frac{1}{h}\left(1 - \frac{1}{Q_{L1}} \right) \right]^2 \right\}},$$

in which F_1 is the pull on one side of the block due to the fundamental component. For $Q_L = 10$, the relative magnitudes of other components are as follows:

h	F_h
1	F_1
3	$F_1/730$
5	$F_1/2324$

For a current source, the quiescent coil currents contain whatever harmonic components are in the source, without attenuation, so that the quiescent harmonic force components are in proportion to the squares of the harmonic current components.

Comparisons of restoring force along one axis of a four-pole magnetic suspension for sinusoidal and square-wave excitations are shown in Figures 5-25 and 5-26. This suspension was connected in accordance with Case (1) for the single-axis suspension, and its action along one axis is essentially the same as the action of the single-axis suspension. (Multiaxis suspensions are discussed in Chapter 7.) The comparisons are made on the basis of equal quiescent power inputs to the suspension. Under that condition the rms value of the applied square wave of voltage is somewhat larger than the rms value of the sinusoid.

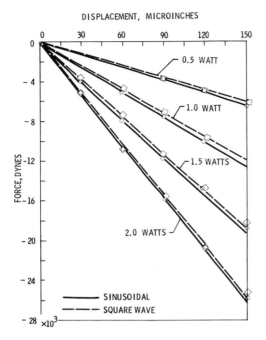

Figure 5-25 Centering force versus displacement along a polar axis of a four-pole suspension with various quiescent power levels for sinusoidal and square-wave applied voltages, Case (1) connection.

Other waveforms, such as triangular or ramp, periodic rectangular, triangular, or half sinusoid pulses, or wave trains of any of these forms, may be used. Wave trains of decaying oscillations resulting from periodic pulses may be used. Various kinds of modulated waveforms may be employed. Most of these alternatives are currently only of academic interest. The triangular wave is lowest in harmonic content. The periodic rectangular pulses are in general easier to obtain than the alternating square wave; also the use of pulses or wave trains allows for time sharing and multiplexing, discussed in Chapter 9. The use of decaying oscillations or modulated waves helps eliminate certain residual effects in magnetic materials, described in Chapter 8, and of importance in active suspensions, discussed in Chapter 9.

Restoring force versus displacement data have been taken along one axis of an eight-pole suspension connected in accordance with Case (1), for 9600-hertz sinusoidal, square, triangular, and square half waves, on an equal rms quiescent-current basis (which aside from differences in core losses is essen-

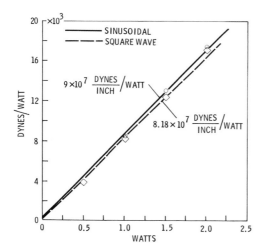

Figure 5-26 Centering force/input power versus quiescent input power for suspension of Figure 5-25.

tially the same as a constant quiescent-power basis), and the data for the various waveforms are hardly distinguishable within experimental error. This circumstance is to be expected from theoretical considerations, because force is proportional to current squared and the fundamental component dominates. The expression for pull on one side of the block when it is centered is

$$F = \frac{\mu A N^2}{g_0^2} \sum_1^h I_h^2 = \frac{LR}{Rg_0} \sum_1^h I_h^2 = \frac{T_L}{g_0} P,$$

in which I_h are the harmonic currents in a coil, however the suspension coils may be connected, and P is the average power delivered to a coil. Hence for fixed quiescent-power input, the quiescent pull on one side of the block is fixed, independent of the harmonic content of the supply and independent of various circuit arrangements that have been considered. As the block is displaced, the attenuation of the various harmonic coil currents in the two circuits is somewhat different, but the net restoring force always can be computed for one frequency at a time, as mentioned at the end of this section. At substantially higher frequencies, the harmonic currents in winding or other stray capacitance might produce significant effects not taken into account in the preceding analyses or observed in the tests.

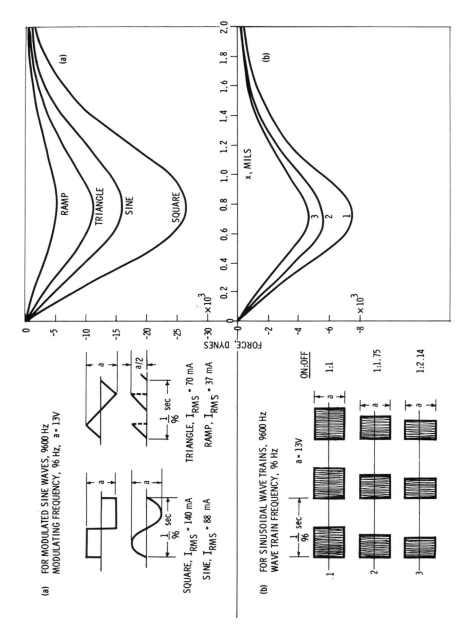

Figure 5-27 Centering force versus displacement along a pole-pair axis of an eight-pole suspension: (a) for various modulated sinusoidal waveforms, (b) for various sinusoidal wave-train pulses, Case (1) connection.

If tests are made with the peak-to-peak voltages of the various waveforms held fixed, then the force curves are quite different. Such curves are shown for various modulated sinusoidal waveforms in Figure 5-27(a), and for various sinusoidal wave-train pulses in Figure 5-27(b). If the modulating ramp, Figure 5-27(a), were *a* high, no zero intervals, or symmetrical about the horizontal axis, *a* peak-to-peak, no zero intervals, the force curve for it would be essentially the same as for the triangular modulation. The force level with sinusoidal wave-train pulses, Figure 5-27(b), is essentially proportional to the fraction of the time that the wave train is on. The force curve for the square-wave modulated sinusoid, Figure 5-27(a), is essentially the same as the force curve for a continuous sinusoid and hence can serve as a basis of comparison for the curves of Figure 5-27(b).

Whereas the comparative data that have been taken are for the connections of Case (1), which has series tuning, the force-displacement curves for the connections that apply to the other cases also are not very sensitive to harmonics in the applied voltage or current. But for the parallel-tuned circuits of Cases (3) and (4) and the bridge circuits of Cases (5), (6), and (8), any direct component in the supply would cause higher peak voltages and corresponding direct current in the coils, which could overheat them and also possibly cause magnetic saturation. Further, forces due to direct-current components in the coils always tend to be destabilizing. Likewise the shunt capacitance in the circuits of Cases (3), (4), and (8) may carry substantial harmonic currents plus any bias voltage from one-sided waves. Formulas already derived can be used for any one harmonic at a time, so that the harmonic contribution to force, stiffness, flux density, coil currents and voltages, capacitor voltages, and bridge signals can be computed. The quality factors Q_L and Q_0 change directly with the order of the harmonic, and

$$Q = Q_L - Q_c = \frac{\omega L}{R} - \frac{1}{2\omega C R}$$

changes more than in proportion to the order of the harmonic.

5-5 Summary

The comparative performances of a magnetic suspension for various connections and various types of excitation depend on the basis of comparison selected. No one mode of operation can of itself be designated as best, or optimum; the comparison must be made on a basis that is suited to the conditions at hand. The tables and plots in this chapter provide means of making such comparative judgments. Such data can be obtained by machine computation ad infinitum; the data here are primarily for illustrative purposes and to aid in the visualization of suspension performance. On the basis of equal power input, excitation by simple nonsinusoidal waves, such as square waves, causes little deterioration in performance compared with performance for sinusoidal excitation. Use of square waves is advantageous because such waves, either full or half wave, are so readily obtained by chopping direct voltage, which requires less auxiliary apparatus and power than the generation of sinusoidal voltage. Whereas a one-sided square wave is easier to obtain by chopping than a square wave symmetrical about the horizontal axis, the one sided wave has the disadvantage of increasing peak voltages and currents, causing direct current where capacitors do not block it, requiring excess magnetic material to avoid saturation, and conceivably causing instability.

6 IDEAL SINGLE-AXIS PASSIVE ELECTRIC SUSPENSIONS

Duals or analogs of the magnetic suspensions discussed in preceding chapters can be constructed that use the pull of the electric field instead of the pull of the magnetic field. Some such suspensions have been built, but for various reasons they have not come into much use in competition with magnetic suspensions.

6-1 Advantages and Disadvantages

A schematic arrangement for a passive electric suspension is shown in Figure 6-1 for voltage-source operation. The circuit for this arrangement does not require grounding of the block. A disadvantage of the electric suspension is the possible need for a flexible lead to the moving member to drain induced charges from it. The principal disadvantage of the electric suspension is the need for high voltage gradients in the gaps to achieve adequate forces; such gradients usually mean relatively high voltages across relatively short gaps. The breakdown voltage of the gaps and the force developed relative to force across an air gap depend on the characteristics of the damping fluid used. If the area dimensions of the capacitor plates and the separation of adjacent capacitors are large compared with the gap length g_0, the leakage capacitance should be relatively small. This condition is in fact necessary to obtain a substantial force without requiring excessive voltage. However, in a small device

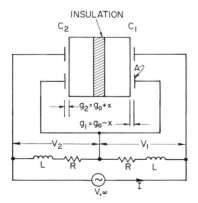

Figure 6-1 Schematic for electric suspension, Cases (3') and (4').

the working capacitance may be quite small, so that the capacitance between leads may be troublesome. Advantages of the electric suspension are the saving of weight, space, and energy loss in the capacitor parts themselves, since the plates may be the suspended body itself or some metal sputtered or plated on to it for the one side, and very thin metal plates for the other side. An insulating coating is desirable to limit short-circuit current, though stops commonly are used that prevent a gap from closing completely. A substantial difference in dielectric constants of the insulating film and the damping fluid can appreciably influence the force level for a short gap. Owing to the fact that the circuitry for the electric suspension must use coils for the tuned circuit or for the bridge connection, the saving in weight, space, and energy dissipation associated with the suspension itself is at least partially lost. However the fact that the location of the coils is more or less arbitrary and that their design can be entirely independent of the suspension structure is a considerable advantage.

The equations for the electric suspension are not developed with the same elaboration as the equations for the magnetic suspension, because a considerable amount of algebraic duplication would be involved and because the electric suspension has not the wide application achieved by the magnetic suspension, though the possibilities of the electric suspension perhaps have not been adequately exploited. The details of performances and comparisons for the various circuit arrangements with electric suspensions can readily be deduced largely by procedures developed for magnetic suspensions in preceding chapters.

6-2 Voltage or Current Source, Parallel Tuning,[38] Cases (3′) and (4′)

For the arrangement of Figure 6-1, the average force acting on the block, in terms of rms magnitudes, is

$$F'_p = \frac{\varepsilon A}{4}\left(\frac{V_1^2}{g_1^2} - \frac{V_2^2}{g_2^2}\right) = \frac{I^2 \varepsilon A}{4}\left(\frac{1}{g_1^2 Y_1^2} - \frac{1}{g_2^2 Y_2^2}\right), \tag{6-1}$$

in which ε is the capacitivity of the damping fluid. The voltage relations are the same in literal form as for the magnetic suspension with parallel tuning, but here the capacitances vary with displacement of the block and the inductances are constant. Hence

$$Y_1 = j\omega\frac{C_1}{2} + G_e + \frac{1}{j\omega L_e} = G_e\left[1 + j\left(D + \frac{D_0 x_n}{1 - x_n}\right)\right]$$

and

$$Y_2 = j\omega \frac{C_2}{2} + G_e + \frac{1}{j\omega L_e} = G_e \left[1 + j \left(D - \frac{D_0 x_n}{1 + x_n} \right) \right],$$

in which

$$C_1 = \frac{C_0}{1 - x_n} + C_l,$$

$$C_2 = \frac{C_0}{1 + x_n} + C_l,$$

$$C_0 = \frac{\varepsilon A}{g_0},$$

$$D = D_c - D_L,$$

$$D_c = \frac{\omega(C_0 + C_l)}{2G_e} = \frac{\omega C}{2G_e},$$

$$D_L = \frac{1}{\omega L_e G_e} = \frac{\omega L}{R} = Q_L,$$

$$D_0 = \frac{\omega C_0}{2G_e},$$

G_e = equivalent shunt conductance of a coil at source frequency,

L_e = equivalent shunt inductance of a coil at source frequency,

and

C_l = leakage capacitance associated with one capacitance, assumed to be constant.

Use of Y_1 and Y_2 to give V_1 and V_2 in terms of V and insertion in Equation (6-1) gives for Case (3'),

$$F_{Vp}' = \frac{V^2 \varepsilon A}{4g_0^2} \frac{(D^2 - D_0 D + 1)x_n}{[D + (D_0 - D)x_n^2]^2 + (1 - x_n^2)^2} = \frac{V^2 \varepsilon A}{4g_0^2} F_n', \tag{6-2}$$

which is identical in form with Equation (3-2) for series tuning with current source, Case (2), or for the untuned bridge circuit with current source, Case

(6), as applied to the magnetic suspension, except for a factor of 4. Here the base for normalizing is

$$F_V' = \frac{V^2 \varepsilon A}{4g_0^2}$$

and is the pull on one side of the block when it is centered. The normalized force curves of Figures 3-2 through 3-4 apply to this case, provided that analogous interpretations are made so that D and Q, D_0 and Q_0 correspond and the ordinate scales are divided by 4. In this circuit, D not only is a measure of the sharpness of tuning but is a measure of the power point at which the circuits operate when the block is centered. Strictly, for actual physical analogy, the coils should be resistanceless and the conductances should relate to the leakages associated with the capacitances, but mathematically the situation would be no different. If significant leakage is associated with the capacitances, the corresponding conductance can be lumped with G_e.

The stiffness of the suspension when the block is centered is

$$\dot{F}_V'\Big|_{x\to 0} = \frac{V^2 \varepsilon A}{4g_0^3} \frac{(D^2 - D_0 D + 1)}{D^2 + 1} = \frac{V^2 \varepsilon A}{4g_0^3} \dot{F}_{n0}'. \tag{6-3}$$

The normalized stiffness curves of Figure (3-9) apply to this case with the same provisos mentioned for Figures (3-2) through (3-4) when interpreted for this electric suspension.

A position signal can be obtained by looking at $V_2 - V_1$, which can be done in effect by placing secondaries on the coils and looking at the difference of the secondary voltages. Voltages V_1 and V_2 are proportional to the respective coil currents, as would be the secondary voltages, so that the signal voltage is

$$V_B = k_v(V_2 - V_1) = k_v V \frac{Y_1 - Y_2}{Y_1 + Y_2}, \tag{6-4}$$

whence

$$|V_n| = \left|\frac{V_B}{V}\right| \approx \frac{k_v D_0 x_n}{\sqrt{D^2 + 1}}, \tag{6-5}$$

for very small displacements.

The quiescent position of the block can be adjusted by use of trimming capacitance c either in admittance Y_1 or in admittance Y_2, the condition being

$$\left|\frac{V_1}{V_2}\right| = \frac{1 - x_{n0}}{1 + x_{n0}} = \left|\frac{Y_2}{Y_1}\right|. \tag{6-6}$$

The result, if the resistances of the tuning coils are equal, which should be readily adjusted, is

$$x_{n0} \approx \pm \frac{D_0 D c_n}{2(D^2 - D_0 D + 1)} \tag{6-7}$$

or

$$c_n \approx \pm \frac{2(D^2 - D_0 D + 1)x_{n0}}{D_0 D}, \tag{6-8}$$

depending on whether c is added in Y_1 or Y_2. Physically negative capacitance c_n of course means that the capacitance would need to be decreased in the admittance to which the adjustment applied, so that adjustment actually must be made by adding capacitance c in the admittance to which positive c_n applies.

Use of Y_1 and Y_2 to give I_{1c} and I_{2c} in terms of I and insertion in Equation (6-1) gives, for Case (4'),

$$F'_{Ip} = \frac{I^2 \varepsilon A}{4G_e^2 g_0^2} \frac{4(D^2 - D_0 D + 1)x_n}{\{[D_0 - (D_0 - D)(1 - x_n)]^2 + (1 - x_n)^2\}} = \frac{I^2 \varepsilon A}{4G_e^2 g_0^2} F'_n, \\ \times \{[D_0 - (D_0 - D)(1 + x_n)]^2 + (1 + x_n)^2\} \tag{6-9}$$

in which the base for normalizing is

$$F'_I = \frac{\varepsilon A}{g_0^2} \frac{I^2}{4G_e^2},$$

which is the pull on one side of the block when the circuit is at resonance. The normalized curves of Figures 2-3 through 2-5 apply to this case, because Equation (6-9) is identical in form with Equation (2-8) for series tuning with voltage source, Case (1), or the tuned bridge circuit with voltage source, Case (7).

The stiffness of the suspension when the block is centered is

$$\dot{F}'_{Ip}\bigg|_{x \to 0} = \frac{I^2 \varepsilon A}{4G_e^2 g_0^3} \frac{4(D^2 - D_0 D + 1)}{(D^2 + 1)^2} = \frac{I^2 \varepsilon A}{4G_e^2 g_0^3} \dot{F}'_{n0}. \tag{6-10}$$

The normalized stiffness curves of Figures 2-12 and 2-13 apply to this case.

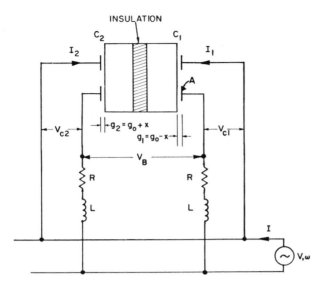

Figure 6-2 Schematic for electric suspension, Cases (1') and (2').

The ratio of stiffness for operation with voltage source and with current source for parallel tuning is

$$\left.\frac{\dot{F}_{Vp}''}{\dot{F}_{Ip}'}\right|_{x\to0} = \frac{V^2 G_e^2}{4I^2}(D^2 + 1).$$ (6-11)

Hence when $I = VG_e/2$, the stiffness when the block is centered is $(D^2 + 1)$ times as much for operation with a voltage source as for operation with a current source. The analogy with Equation (3-9) is complete, because for the electric suspension with parallel tuning $G_e/2$ is the conductance seen by the current source at resonance, whereas for the magnetic suspension R is the resistance seen by the voltage source at resonance.

The equation for position signals taken in the same manner as for operation with a voltage source becomes

$$V_B = k_I I\left(\frac{1}{Y_2} - \frac{1}{Y_1}\right),$$ (6-12)

and for very small displacements

$$|V_n| = \left|\frac{V_B G_e}{I}\right| \approx \frac{2k_I D_0 x_n}{D^2 + 1}.$$ (6-13)

The relations for displacing the quiescent position of the block are the same as given in Equations (6-7) and (6-8).

6-3 Voltage or Current Source, Series Tuning,[41] Cases (1′) and (2′)

For the arrangement of Figure 6-2, the average force acting on the block, in terms of rms magnitudes, is

$$F'_s = \frac{\varepsilon A}{4}\left(\frac{V_{c1}^2}{g_1^2} - \frac{V_{c2}^2}{g_2^2}\right) = \frac{\varepsilon A}{\omega^2}\left(\frac{I_1^2}{g_1^2 C_1^2} - \frac{I_2^2}{g_2^2 C_2^2}\right), \tag{6-14}$$

in which V_{c1} and V_{c2} are the voltages across the pairs of capacitors in series at each end. This circuit is not an exact analog of any of the circuits for the magnetic suspension, because the coil resistance cannot be represented as an equivalent constant conductance across the associated capacitance, but the association can be made mathematically. The impedances may be written

$$Z_1 = \frac{1}{G_e + \dfrac{1}{j\omega L_e}} + \frac{2(1 - x_n)}{j\omega[C - (C - C_0)x_n]}$$

$$= \frac{1}{G_e}\left\{\frac{1}{D_L^2 + 1} + j\left[\frac{D_L}{D_L^2 + 1} - \frac{1 - x_n}{D_c - (D_c - D_0)x_n}\right]\right\}$$

and

$$Z_2 = \frac{1}{G_e + \dfrac{1}{j\omega L_e}} + \frac{2(1 + x_n)}{j\omega[C + (C - C_0)x_n]}$$

$$= \frac{1}{G_e}\left\{\frac{1}{D_L^2 + 1} + j\left[\frac{D_L}{D_L^2 + 1} - \frac{1 + x_n}{D_c + (D_c - D_0)x_n}\right]\right\}.$$

Use of these impedances and their capacitive parts to obtain V_{c1} and V_{c2} in terms of V gives, for Case (1′),

$$F'_{Vs} = \frac{V^2 \varepsilon A}{4 g_0^2} \frac{4[(D_c - D)^2 + 1][D^2 - D_0 D + 1]x_n}{\{[D_0 - (D_0 - D)(1 - x_n)]^2 + (1 - x_n)^2\}} = \frac{V^2 \varepsilon A}{4 g_0^2} F'_n,$$
$$\times \{[D_0 - (D_0 - D)(1 + x_n)]^2 + (1 + x_n)^2\}$$

$$\tag{6-15}$$

the normalized part of which can be obtained by multiplying Equation (2-8), for Case (1), or the ordinates of Figures 2-3 through 2-5 by $(D_L^2 + 1)$. This coefficient is left in terms of D_c and D in Equation (6-15) simply for conformity with other equations. In making the translation from Equation (2-8) or from Figures 2-3 through 2-5, D_L of course cannot be chosen independently but must correspond to the values of D_c and D for which F'_{Vs} is desired. The stiffness at zero displacement is

$$\dot{F}_{Vs}\bigg|_{x\to 0} = \frac{V^2 \varepsilon A}{4g_0^3} \frac{4[(D_c - D)^2 + 1][D^2 - D_0 D + 1]}{(D^2 + 1)^2} = \frac{V^2 \varepsilon A}{4g_0^3} \dot{F}'_{n0}. \qquad (6\text{-}16)$$

In Equation (6-16) the quantity $(D_c - D)^2 \gg 1$, so that with very little error Equation (3-28) for Case (4) divided by 4 can be used, in which Q_L is analogous to D_c, and Figure 3-22 can be used, with the ordinate scale for Case (8) and recognition that the restriction $D_c/D_0 = \frac{3}{2}$ must be accepted. In all these equations involving D_c and D_0, one must remember that they do not apply to the capacitors themselves but involve the resistances of the coils. The quality factors of the capacitors themselves should be extremely high.

Position signals can be obtained from the bridge circuit as indicated in Figure 6-2,

$$V_n = \frac{V_B}{V} = \frac{1 + jD_L}{1 + j\left[D_L - \dfrac{(D_L^2 + 1)(1 + x_n)}{D_c + (D_c - D_0)x_n}\right]} - \frac{1 + jD_L}{1 + j\left[D_L - \dfrac{(D_L^2 + 1)(1 - x_n)}{D_c - (D_c - D_0)x_n}\right]},$$

$$\qquad (6\text{-}17)$$

which for very small displacements reduces to

$$|V_n| \approx \frac{2D_0\sqrt{D_L^2 + 1}}{D^2 + 1}x_n = \frac{2D_0\sqrt{(D_c - D)^2 + 1}}{D^2 + 1}x_n. \qquad (6\text{-}18)$$

Adjustment of quiescent position of the block can be accomplished by use of trimming capacitance c across the working capacitance on one end or the other. If c is incorporated into Z_1,

$$X_{c1} \approx -\frac{2(1 - x_{n0})}{\omega[C(1 - x_{n0}) + C_0(c_n + x_{n0})]} \approx -\frac{1}{G_e}\left[\frac{1}{D_c + D_0(c_n + x_{n0})}\right],$$

for very small values of x_{n0}. Similarly, if c is incorporated into Z_2,

$$X_{c2} \approx -\frac{2(1 + x_{n0})}{\omega[C(1 + x_{n0}) + C_0(c_n - x_{n0})]} \approx -\frac{1}{G_e}\left[\frac{1}{D_c + D_0(c_n - x_{n0})}\right].$$

Hence,

$$Z_1 \approx \frac{1}{G_e} \left\{ \frac{1}{D_L^2 + 1} + j \left[\frac{D_L}{D_L^2 + 1} - \frac{1}{D_c + D_0(c_n + x_{n0})} \right] \right\}$$

and

$$Z_2 \approx \frac{1}{G_e} \left\{ \frac{1}{D_L^2 + 1} + j \left[\frac{D_L}{D_L^2 + 1} - \frac{1}{D_c - D_0 x_{n0}} \right] \right\},$$

or

$$Z_1 \approx \frac{1}{G_e} \left\{ \frac{1}{D_L^2 + 1} + j \left[\frac{D_L}{D_L^2 + 1} - \frac{1}{D_c + D_0 x_{n0}} \right] \right\}$$

and

$$Z_2 \approx \frac{1}{G_e} \left\{ \frac{1}{D_L^2 + 1} + j \left[\frac{D_L}{D_L^2 + 1} - \frac{1}{D_c + D_0(c_n - x_n)} \right] \right\}.$$

The condition for balance is

$$\frac{V_{c1}}{V_{c2}} = \frac{1 - x_{n0}}{1 + x_{n0}} = \left| \frac{X_{c1}}{Z_1} \frac{Z_2}{X_{c2}} \right|, \tag{6-19}$$

evaluation of which gives

$$c_n \approx \mp \frac{2(D^2 - D_0 D + 1) x_{n0}}{D_0 D} \tag{6-20}$$

or

$$x_{n0} \approx \mp \frac{D_0 D c_n}{2(D^2 - D_0 D + 1)} \tag{6-21}$$

for very small displacements, depending on whether c is incorporated in Z_1 or Z_2, the same as for Cases (3′) and (4′), though not obviously.

When a current source is used and Z_1 and Z_2 are used to obtain currents I_1 and I_2 in terms of I, Equation (6-14) gives, for Case (2′),

$$F'_{Is} = \frac{I^2 \varepsilon A}{4 G_e^2 g_0^2} \frac{[D^2 - D_0 D + 1] x_n}{[D_c D + (D_c - D_0)(D_0 - D) x_n^2]^2 + [D_c - (D_c - D_0) x_n^2]^2}$$

$$= \frac{I^2 \varepsilon A}{4 G_e^2 g_0^2} F'_n, \tag{6-22}$$

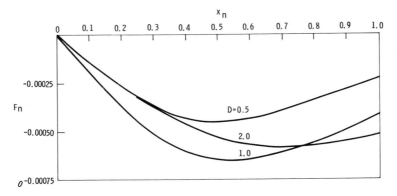

Figure 6-3 Normalized centering force versus normalized displacement, Case (2'); $D_0 = 6$.

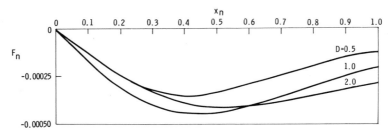

Figure 6-4 Normalized centering force versus normalized displacement, Case (2'); $D_0 = 10$.

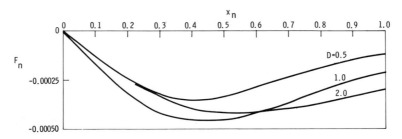

Figure 6-5 Normalized centering force versus normalized displacement, Case (2'); $D_0 = 16$.

which is plotted in Figures 6-3 through 6-5 and in the absence of leakage capacitance becomes quite simple and perfectly linear. It has no magnetic counterpart. In normalized form, these forces are comparable to the normalized forces of Case (3) and hence are relatively small.

The stiffness of the suspension when the block is centered is

$$\dot{F}'_{Is}\bigg|_{x\to 0} = \frac{I^2\varepsilon A}{4G_e^2 g_0^3}\frac{D^2 - D_0 D + 1}{D_c^2(D^2 + 1)} = \frac{I^2\varepsilon A}{4G_e^2 g_0^3}\dot{F}'_{no}. \tag{6-23}$$

The normalized stiffness $\dot{F}'_{no}$ can be obtained by dividing Equation (3-8) for Case (2) or the ordinates of Figure 3-9 by $4D_c^2$.

Stiffness relations for the various tuned electric suspensions are

$$\frac{\dot{F}'_{Vs}}{\dot{F}'_{Is}}\bigg|_{x\to 0} = \frac{V^2 G_e^2}{I^2}\frac{4D_c^2[(D_c - D)^2 + 1]}{D^2 + 1}, \tag{6-24}$$

$$\frac{\dot{F}'_{Vp}}{\dot{F}'_{Is}}\bigg|_{x\to 0} = \frac{V^2 G_e^2 D_c^2}{I^2}, \tag{6-25}$$

and

$$\frac{\dot{F}'_{Ip}}{\dot{F}'_{Is}}\bigg|_{x\to 0} = \frac{4D_c^2}{D^2 + 1}. \tag{6-26}$$

In terms of current I, the position signal for very small displacement is

$$|V_n| = \left|\frac{V_B G_e}{I}\right| = \frac{D_0 x_n}{D_c\sqrt{D^2 + 1}}. \tag{6-27}$$

The relations for displacing the quiescent position of the block by capacitance trimming are the same as given in Equation (6-20) and Equation (6-21).

6-4 Voltage or Current Source, Untuned Bridge Circuit, Cases (5′) and (6′)

In Figure 6-6, the capacitors at one end of the block, Figure 6-1, are assumed to be in opposite branches of the bridge circuit, and the capacitors at the other end are assumed to be in the other opposite branches, with $C_2 = C_3$ and $C_1 = C_4$. But this bridge-circuit arrangement is physically impossible with the construction shown in Figure 6-1. Hence to make possible the bridge-circuit connections of Figure 6-6, the structure is reconstructed as shown in Figure 6-7. This arrangement requires the coil to be connected to the block by flexible leads unless the coil can ride with the block, which is conceivable.

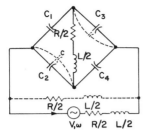

Figure 6-6 Bridge-circuit connection for Case (5′), (6′), (7′), or (8′).

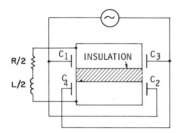

Figure 6-7 Schematic for electric suspension, Cases (5′), (6′), (7′), or (8′).

If position signals are desired from the coil or if trimming capacitors are used, flexible leads of course must be attached to the block.

First, no impedance is supposed to be associated with the source, as indicated by $R/2$ and $L/2$ shown dotted, and trimming capacitances c are supposed to be absent. Though strictly the circuit is not tuned, the relation

$$D = \frac{\omega C}{2G_e} - \frac{1}{2\omega L_e G_e} = D_c - D_L$$

still is analytically useful. Solution of the bridge circuit for its currents in terms of source voltage and impedances gives, for Case (5′)

$$F'_{VB} = \frac{V^2 \varepsilon A}{4g_0^2} \frac{4(D^2 - D_0 D + 1)x_n}{[D + (D_0 - D)x_n^2]^2 + (1 - x_n^2)^2} = \frac{V^2 \varepsilon A}{4g_0^2} F'_n. \tag{6-28}$$

This equation is the same as Equation (6-2) for Case (3′) multiplied by 4 and has the same form as Equation (3-2), Case (2), and hence Figures 3-2 through 3-4 represent the normalized part. The stiffness at zero displacement is

$$\dot{F}'_{VB}\bigg|_{x\to 0} = \frac{V^2\varepsilon A}{4g_0^3}\frac{4(D^2 - D_0 D + 1)}{D^2 + 1} = \frac{V^2\varepsilon A}{4g_0^3}\dot{F}'_{n0}, \tag{6-29}$$

which is the same as Equation (6-3) multiplied by 4 and has the same form as Equation (3-7), so that Figure 3-9 can represent the normalized part of Equation (6-29) also.

The comparative stiffnesses for operation with this bridge circuit and the tuned circuits are

$$\frac{\dot{F}'_{VB}}{\dot{F}'_{Vp}}\bigg|_{x\to 0} = 4, \tag{6-30}$$

$$\frac{\dot{F}'_{VB}}{\dot{F}'_{Ip}}\bigg|_{x\to 0} = \frac{V^2 G_e^2}{I^2}(D^2 + 1), \tag{6-31}$$

$$\frac{\dot{F}'_{VB}}{\dot{F}'_{Vs}}\bigg|_{x\to 0} = \frac{D^2 + 1}{(D_c - D)^2 + 1}, \tag{6-32}$$

and

$$\frac{\dot{F}'_{VB}}{\dot{F}'_{Is}}\bigg|_{x\to 0} = \frac{V^2 G_e^2}{I^2}4D_c^2. \tag{6-33}$$

Solution of the bridge circuit for its currents in terms of source current and impedances gives for F'_{IB}, Case (6'), the same result as for series tuning when a current source is used, Equation (6-22), Case (2'), but it has no magnetic counterpart.

The voltage across the coil can be taken as the position signal:

$$V_B = (I_1 - I_2)\frac{R + j\omega L}{2} = \frac{I_1 - I_2}{2G_e(1 - jD_L)}. \tag{6-34}$$

For a voltage source, Case (5'), the result is

$$|V_n| = \left|\frac{V_B}{V}\right| = \frac{D_0 x_n}{\sqrt{D^2 + 1}}, \tag{6-35}$$

and for a current source, Case (6'), the result is

$$|V_n| = \left|\frac{V_B G_e}{I}\right| = \frac{D_0 x_n}{2D_c\sqrt{D^2 + 1}} \tag{6-36}$$

for very small displacements.

Positioning of the block may be achieved by shunting trimming capacitance c around branch 2 and around branch 3 or around branch 1 and branch 4, depending on the desired direction of displacement. For example, if

$$\left|\frac{V_{c1}}{V_{c2}}\right| = \frac{1 - x_{n0}}{1 + x_{n0}} = \left|\frac{I_1 X_{c1}}{I_2 X_{c2}}\right|,$$

then

$$c_n \approx \frac{2[D^2 - D_0 D + 1]x_{n0}}{D_0 D} \tag{6-37}$$

or

$$x_{n0} \approx \frac{D_0 D c_n}{2[D^2 - D_0 D + 1]} \tag{6-38}$$

for very small displacements, if c is incorporated into branches 1 and 4. If c is incorporated into branches 2 and 3, the relative signs of c_n and x_{n0} reverse. These relations are the same as for Cases (1′) through (4′), but not obviously.

6-5 Voltage or Current Source, Tuned Bridge Circuit, Cases (7′) and (8′)

If now resistance $R/2$ and inductance $L/2$, shown dotted in Figure 6-6, are inserted in series with voltage source V, the definition of Q is satisfied as an indication of sharpness of tuning of the bridge circuits as a whole and as an indication of power-point setting when the block is centered, but D does not have that significance. Nevertheless, for conformity and comparison with the solutions for the other circuits used with the electric suspension, D rather than Q is chosen. Solution of this bridge circuit for branch currents in terms of its impedances and voltage source gives, for F'_{VBT}, Case (7′), the same result as for Case (1′), Equation (6-15). The stiffness is given by Equation (6-16) and can be represented with good approximation by Figure 3-22 for Case (8). The tuning can be accomplished also by placing the series combination of resistance $R/2$ and inductance $L/2$ in parallel with current source I, Case (8′), which makes no difference in the behavior of the remainder of the bridge circuit. The force and stiffness equations in terms of current source I can be obtained by making the substitution

$$V^2 = \frac{(R^2 + \omega^2 L^2)I^2}{4} = \frac{I^2}{4G_e^2(D_L^2 + 1)},$$

which gives, Case (8'),

$$
F'_{IBT} = \frac{I^2 \varepsilon A}{4G_e^2 g_0^2} \frac{(D^2 - D_0 D + 1)x_n}{\{[D_0 - (D_0 - D)(1 - x_n)]^2 + (1 - x_n)^2\} \times \{[D_0 - (D_0 - D)(1 + x_n)]^2 + (1 + x_n)^2\}} = \frac{I^2 \varepsilon A}{4G_e^2 g_0^2} F'_n,
$$

(6-39)

which is Equation (6-9) for Case (4') divided by 4. Likewise,

$$
\dot{F}_{IBT}\bigg|_{x\to 0} = \frac{I^2 \varepsilon A}{4G_e^2 g_0^3} \frac{(D^2 - D_0 D + 1)}{(D^2 + 1)^2} = \frac{I^2 \varepsilon A}{4G_e^2 g_0^3} \dot{F}'_{n0},
$$

(6-40)

which is Equation (6-10) divided by 4.

The position signal, derived from Equation (6-34), is

$$
|V_n| = \left|\frac{V_B}{V}\right| = \frac{D_0 \sqrt{(D_c - D)^2 + 1}}{D^2 + 1} x_n
$$

(6-41)

for very small displacements, and the relations for trimming capacitance are given by Equation (6-37) and Equation (6-38).

6-6 Comparison of Performances for the Various Connections

The series-tuned magnetic-suspension circuit and the parallel-tuned electric-suspension circuit can be viewed as true duals, because with inductance fixed the coils in the parallel-tuned circuit can be represented by their equivalent shunt conductances and inductances. However, the series-tuned electric-suspension circuit and the parallel-tuned magnetic-suspension circuit cannot be viewed as true duals, because with inductance a function of displacement of the block the coils in the parallel-tuned circuit cannot be represented by fixed equivalent shunt conductances and inductances. Therefore, for Cases (1') or (7') and (2') or (6'), no exactly corresponding magnetic suspension behavior exists. But exactly the same interrelations exist among the eight electric suspensions as exist among the eight magnetic suspensions, as summarized in Section 4-8. These circumstances are now tabulated. First for all passive magnetic- and electric-suspension circuits considered:*

$$
\begin{array}{llll}
F'_{Vp} & [\text{Case (3')}] & \leftrightarrow & F_{Is}/4 & \quad [\text{Case (2) or (6)}], \\
F'_{Ip} & [\text{Case (4')}] & \leftrightarrow & F_{Vs} & \quad [\text{Case (1) or (7)}],
\end{array}
$$

* All normalized forces.

F'_{Vs} [Case (1')] $\leftrightarrow$ $(D_L^2 + 1)F_{Vs}$ [Case (1) or (7)],
F'_{Is} [Case (2')] $\leftrightarrow$ no magnetic counterpart,
F'_{VB} [Case (5')] $\leftrightarrow$ F_{Is} [Case (2) or (6)],
F'_{IB} [Case (6')] $\leftrightarrow$ no magnetic counterpart,
F'_{VBT} [Case (7')] $\leftrightarrow$ $(D_L^2 + 1)F_{Vs}$ [Case (1) or (7)],
F'_{IBT} [Case (8')] $\leftrightarrow$ $F_{Vs}/4$ [Case (1) or (7)].

Then within the electric-suspension circuits considered:

F'_{VB} [Case (5')] $=$ $4F'_{Vp}$ [Case (3')], equal V,
F'_{IB} [Case (6')] $=$ F'_{Is} [Case (2')], equal I,
F'_{VBT} [Case (7')] $=$ F'_{Vs} [Case (1')], equal V,
F'_{IBT} [Case (8')] $=$ $F'_{Ip}/4$ [Case (4')], equal I.

For the tabulation of electric-suspension circuits against magnetic-suspension circuits, the comparison of actual forces would require that the normalizing coefficients be taken into account. The same tabulation (also restricted to normalized values) applies for stiffnesses, with two exceptions. For Cases (1') and (7'), since $D_L^2 + 1 = (D_c - D)^2 + 1$, the stiffness really can be written

$$\dot{F}'_{n0} \text{ [Case (1')]} = \dot{F}_{n0}/4 \text{ [Case (4)]} + \dot{F}_{n0} \text{ [Case (1)]},$$

but since $(D_c - D)^2 \gg 1$, $\dot{F}_{n0}$ [Case (1)] is negligible compared with $\dot{F}_{n0}$ [Case (4)]. For Cases (2') and (6'), though no magnetic correspondence exists for the force equations, the initial stiffnesses reduce to the relation

$$\dot{F}'_{n0} \text{ [Case (2')]} = \dot{F}_{n0}/4D_c^2 \text{ [Case (2)]}.$$

Significant comparisons could be made among the performances of the electric suspensions with various circuit connections on bases other than equal voltage or current excitations, or $V = RI$, just as was done for the magnetic suspensions: Equal electric displacement limits, equal voltage limits for coils or capacitors, equal maximum forces, equal quiescent stiffnesses, and so on could be taken as bases, but these comparisons would be largely mathematical duplications of work already done or can be readily made by utilizing relations derived herein. But the response to nonsinusoidal excitation of the electric suspension is somewhat different from the response of the magnetic suspension and is noted in the following section.

6-7 Nonsinusoidal Excitation

For series tuning, Cases (1') and (2'), the average quiescent force response of electric suspensions to harmonics in the driving voltage or current should be the same, relatively, as the average quiescent force response of magnetic suspensions. A direct component in a voltage source appears across each capacitor pair and in addition to increasing the resultant voltage there conceivably could cause instability. Any force components due to direct voltage components across the capacitors tend to be destabilizing. A direct component in a current source is incompatible with a circuit containing a series capacitor. For parallel tuning with a voltage source, Case (3'), with zero displacement, the voltage across the capacitors has the same harmonic content as the source, so that the quiescent force components are attenuated in accordance with the squares of the amplitudes of the harmonic voltage components. Of course, the harmonic currents in the capacitors are larger in comparison with the fundamental current but can be harmful only to the extent that they may cause excessive dielectric loss. For parallel tuning with a current source, Case (4'), the quiescent voltage harmonic across a pair of capacitors at one end of the block is

$$V_{ch} = \frac{2I_{ch}}{jh\omega_1 C} = \frac{I_h}{\dfrac{1}{R + jh\omega_1 L} + \dfrac{jh\omega_1 C}{2}} = \frac{2(R + jh\omega_1 L)I_h}{2 + (R + jh\omega_1 L)jh\omega_1 C}$$

or

$$V_{ch}^2 = \frac{R^2(1 + h^2 Q_{L1}^2)I_h^2}{\left(\dfrac{h}{Q_{c1}}\right)^2 \left[1 + \left(\dfrac{Q_{c1}}{h} - hQ_{L1}\right)^2\right]},$$

in which

$$Q_{c1} = \frac{1}{\omega_1 CR/2},$$

$$Q_{L1} = \frac{\omega_1 L}{R},$$

and I_h is the harmonic current in the source. A direct component in the source, for either Case (3') or (4'), gives voltage $V_d = 2RI_d$ across each capacitor pair and in addition to increasing the resultant voltage there conceivably could cause instability.

For the bridge circuits, tuned or untuned, with voltage source, Cases (5') and (7'), a direct component in the source appears across each side of the bridge circuit; and for the tuned bridge circuit with current source having a direct component, a direct voltage $V_d = 2RI_d$ appears across each side of the bridge circuit. These direct voltage components not only increase the voltages across the capacitors but conceivably could cause instability. For the untuned bridge circuit with current source, a direct component in the source is incompatible with a capacitance bridge circuit. The leakage conductances across the capacitors may be an important factor in influencing the distribution of direct voltage across the capacitors. As for the magnetic suspension, the actual harmonic force component for any block displacement can be computed for any voltage- or current-source harmonic by recognizing that for harmonic of order h,

$$D_{Lh} = \frac{h\omega_1 L}{R} = hQ_{L1},$$

$$D_{ch} = \frac{h\omega_1 C}{2R}(R^2 + h^2\omega_1^2 L^2) = \frac{h}{Q_{c1}}(1 + h^2 Q_{L1}^2),$$

$$D_{0h} = \frac{h\omega_1 C_0}{2R}(R^2 + h^2\omega_1^2 L^2) = \frac{h}{Q_{c01}}(1 + h^2 Q_{L1}^2),$$

and

$$D_h = D_{ch} - D_{Lh} = \frac{h}{Q_{c1}}(1 + h^2 Q_{L1}^2) - hQ_{L1}.$$

For all cases the harmonic force components decrease very substantially with increase in order of the harmonic.

6-8 Combined Magnetic and Electric Suspensions[42]

Instead of utilizing either the magnetic field or the electric field to exert the pull for a suspension, with tuning accomplished by means of an external static capacitor or inductor, pull may be obtained from magnetic and electric fields simultaneously, as illustrated schematically in Figure 6-8; and circuit connections may be such that the inductive and capacitive parts of the structure are tuned against each other, either in series or in parallel, but the idea is not applicable to the bridge circuit. Theoretically, a much stiffer suspension can result than is possible by tuning by means of an external static component, but peak force occurs at much smaller displacement. The reason for these results is that both inductive and capacitive reactances change with change

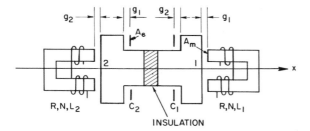

Figure 6-8 Schematic for single-axis combined electric and magnetic suspension.

of displacement and in opposite directions; therefore, the coil currents and the capacitor voltages change much more rapidly than they do when only inductance or only capacitance changes. For example, for series tuning with adjustment for half-power-point operation with the block centered, the stiffness of the combined suspension theoretically may be around 5 to 10 times as much as the stiffness of the magnetic part acting alone, depending on the proportions of leakage inductances and capacitances; and the peak forces theoretically may be about 2 to 4 times as much, but at about $\frac{1}{2}$ to $\frac{3}{4}$ as much displacement. However, to achieve this tuning the capacitor plate area would need to be many times larger than possibly can be built onto a small suspension. In other words, for the combined device, both the inductance and the capacitance are limited by the geometry of the device with which they are associated. For suspending a large device, such as a telescope in a spaceship, for example, or for high source frequencies unattended by excessive core or dielectric losses, the idea might become practical.

6-9 Summary

The force-displacement expressions for the passive electric suspension in general have counterparts in the force-displacement expressions for the passive magnetic suspension. The reason why the analogy is not complete is that although the series-tuned magnetic suspension circuits, Cases (1) and (2), and the parallel-tuned electric suspension circuits, Cases (3') and (4'), can be viewed as true duals, the series-tuned electric suspension circuits, Cases (1') and (2'), and the parallel-tuned magnetic suspension circuits, Cases (3) and (4), cannot be so viewed. To achieve the dual, the coil in the parallel-tuned circuit must be represented in terms of equivalent shunt conductance and inductance; and if the inductance is a function of block position, the equivalent parameters

are not constant with change of gap length. Further, the equivalent parameters are functions of frequency even if the coil inductance is fixed, which is
a consideration when analyzing response to harmonics in the source. But
many of the quantities of interest in the analysis of passive electric suspensions
can be obtained directly from equations and plots developed for magnetic
suspensions merely by reinterpretation of symbols.

The electric suspension has the advantage of small weight and small energy
loss in the device itself; this advantage is somewhat lost owing to the fact that
associated tuning coils are needed. A principal disadvantage is the need of
very high voltage gradients within the gaps and high voltages across them to
achieve adequate force. The performance of the device may be impaired by
induced charge on the floating member if it is not grounded by a flexible lead
and by the tendency of capacitance between leads or elsewhere to short-circuit
the source.

7 IDEAL MULTIAXIS SUSPENSIONS

The ideas of the passive single-axis suspension that have been considerably elaborated in preceding chapters can be extended to control the position of a body along two or three orthogonal axes. For small displacements, the performance along any axis is nearly the same as has been shown for single-axis suspensions, but for large deflections, within the confines of gap clearances, the couplings among the coils for the various axes cause departures from the characteristics of a single-axis suspension. The algebra that specifies these characteristics is too voluminous for hand solution, but a number of solutions have been made by digital computer.

7-1 Extension to Two Orthogonal Axes

To use a magnetic suspension for centering when displacement may occur in any direction in a plane, either four or eight magnetic poles generally are used, as illustrated by Figures 7-1 and 7-2 for application to a cylinder.

For ideal magnetic material, that is, for material having no losses and infinite permeability, the stiffness of the four-pole suspension, Figure 7-1, theoretically is the same in all directions at zero displacement and in fact is identical in form with the stiffness of the single-axis suspension with series tuning, Case (1). When the cylinder is displaced radially, the expressions governing the x- and y-axis restoring forces differ somewhat from the corresponding expressions for the single-axis suspension, depending on the direction of displacement, and are quite laborious to derive rigorously, because each coil is coupled to every other coil. However, some computer solutions are available for this arrangement.

More commonly, an eight-pole structure is used, as shown in Figure 7-2, for which the coils are connected in pairs. For example, for the series-tuning connection, the coils on poles 8 and 1 are connected series aiding magnetically and in series with a capacitor; likewise coils on poles 2 and 3, 4 and 5, and 6 and 7 are paired. A pair of coils, being on the same magnetic circuit, has about four times the self-inductance of one coil and thus greatly reduces the amount of tuning capacitance required. The original reason for this arrangement was to combine the magnetic suspension with a Microsyn[6]* (which cannot be done with a four-pole structure) as illustrated in Figure 7-3, so that in the suspension of a floated instrument, as discussed in some detail in

---

* Electromechanical sensors and actuators will be the subject of another monograph.

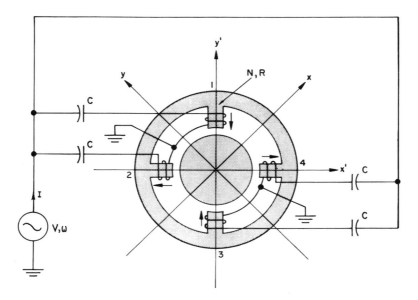

Figure 7-1 Schematic for four-pole cylindrical magnetic suspension.

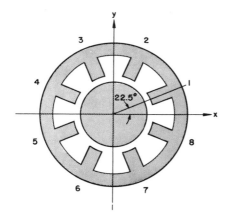

Figure 7-2 Structure for eight-pole cylindrical magnetic suspension.

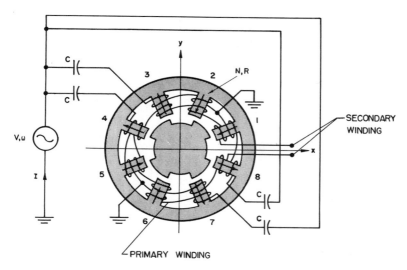

Figure 7-3 Combined eight-pole Microsyn and magnetic suspension with common magnetic circuits and common windings.

Chapter 11, the windings and structure of a magnetic suspension and Microsyn torquer could be combined at one end of the float, and the windings and structure of a magnetic suspension and Microsyn signal generator could be combined at the other end of the float, thus achieving substantial economy of space and weight. The auxiliary winding shown in Figure 7-3 gives a voltage output indicative of angular departure of the rotors from the neutral position shown or, if energized at the same frequency as the main winding, causes a torque to be exerted on the rotor. However, the economy achieved by the combination instruments was at the expense of some sacrifice in performance as compared with the use of separate components for the respective functions. Currently, therefore, the functions are separated electrically but somewhat combined structurally in the Ducosyn,* as illustrated in Figure 7-4, with the suspension inside, Microsyn outside.

For this combination an eight-pole suspension nevertheless is used because, for cylindrical suspensions generally, analysis and experience have shown that it makes more favorable use of iron and copper than the four-pole suspension. The pairing of the coils reduces the number of capacitors required from eight to four and requires only about one-fourth as much total capaci-

* See preceding footnote.

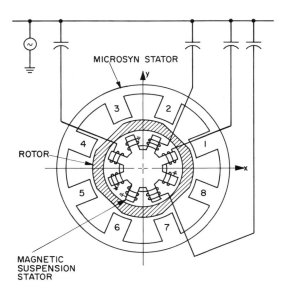

Figure 7-4 Combined eight-pole Microsyn and magnetic suspension with separate magnetic circuits and separate windings, a Ducosyn.

tance as would the tuning of each coil in a separate circuit. The four-pole suspension requires that the magnetic poles alternate in sequence NSNS (otherwise displacement along one axis would upset the equilibrium along the other axis) and is a reason why the combination cannot be made with the Microsyn signal or torque generator, which requires the NNSS sequence. However, the eight-pole magnetic suspension when not combined with the Microsyn can use the alternating sequence of magnetic poles and thus require essentially half as much material in the yoke arcs of the stator and essentially half the flux density in the rotor as when the sequence NNSSNNSS is used, but at some loss of stiffness and considerable increase of leakage fluxes. However, these disadvantages are generally overriding, and the NNSSNNSS sequence is hence used. The larger number of poles makes restoring force and stiffness somewhat more uniform with respect to direction of disturbance and gives shorter lengths of magnetic circuit in the yoke arcs of the stator and in the rotor, per operating coil.

The expressions governing the behavior of the eight-pole magnetic suspension are of course even more laborious to derive than the expressions for the four-pole suspension, owing to the mutual couplings among the coils, but some computer solutions are available for the eight-pole suspension also.[15]

7-2 Derivation of Equations for the Four-Pole Cylindrical Magnetic Suspension[4,5,7,8,11,14,15,43,44]

This suspension does not offer a possibility of approximate solution by neglecting the couplings of the various coils. Equations for rigorous solution within the assumptions of ideal magnetic core materials, as previously mentioned, and parallel-faced gaps too short compared with their cross-sectional dimensions to require fringing corrections are therefore established and normalized in a manner convenient for computer solutions. The x' and y' axes give simpler solutions for this case, rather than the x and y axes, which give simpler solutions when coils can be paired in series, as described for the eight-pole suspension. As an example, equations are written for the series-tuned connection with voltage source, Case (1).

The magnetic-circuit equations, Figure 7-1, are

$$NI_1 + NI_2 = \mathcal{R}_1\Phi_1 + \mathcal{R}_2\Phi_2, \tag{7-1}$$

$$NI_2 + NI_3 = \mathcal{R}_2\Phi_2 + \mathcal{R}_3\Phi_3, \tag{7-2}$$

$$NI_3 + NI_4 = \mathcal{R}_3\Phi_3 + \mathcal{R}_4\Phi_4, \tag{7-3}$$

$$NI_4 + NI_1 = \mathcal{R}_4\Phi_4 + \mathcal{R}_1\Phi_1, \tag{7-4}$$

and

$$\Phi_1 + \Phi_3 = \Phi_2 + \Phi_4. \tag{7-5}$$

Here the $\mathcal{R}$s represent the reluctances of the respective gaps. Of Equations (7-1) through (7-4) only three are independent. The electric-circuit equations are

$$V = \left(R + j\omega L_\ell + \frac{1}{j\omega C}\right)I_1 + j\omega N\Phi_1, \tag{7-6}$$

$$V = \left(R + j\omega L_\ell + \frac{1}{j\omega C}\right)I_2 + j\omega N\Phi_2, \tag{7-7}$$

$$V = \left(R + j\omega L_\ell + \frac{1}{j\omega C}\right)I_3 + j\omega N\Phi_3, \tag{7-8}$$

and

$$V = \left(R + j\omega L_\ell + \frac{1}{j\omega C}\right)I_4 + j\omega N\Phi_4. \tag{7-9}$$

The component forces along the respective axes are

$$F_{x'} = \frac{1}{2\mu A}(\Phi_4^2 - \Phi_2^2) \tag{7-10}$$

and

$$F_{y'} = \frac{1}{2\mu A}(\Phi_1^2 - \Phi_3^2). \tag{7-11}$$

These equations can be generalized and adapted to forms more favorable to simultaneous solution by means of a computer by introduction of Q_0 and Q, where, as usual,

$$Q_0 = \frac{\omega N \Phi_0}{R I_0} = \frac{\omega L_0}{R}$$

and

$$Q = \frac{\omega L_\ell}{R} - \frac{1}{\omega C R} + \frac{\omega N \Phi_0}{R I_0} = \frac{\omega (L_\ell + L_0)}{R} - \frac{1}{\omega C R},$$

in which Φ_0 and I_0 are, respectively, the polar flux and the coil current with the cylinder centered, and by the introduction of normalized incremental quantities resulting from departure of the cylinder from its centered position,

$$I_1 = I_0\left(1 + \frac{i_1}{I_0}\right) = I_0(1 + i_{n1}),$$

$$I_2 = I_0\left(1 + \frac{i_2}{I_0}\right) = I_0(1 + i_{n2}),$$

$$I_3 = I_0\left(1 + \frac{i_3}{I_0}\right) = I_0(1 + i_{n3}),$$

and

$$I_4 = I_0\left(1 + \frac{i_4}{I_0}\right) = I_0(1 + i_{n4});$$

$$\mathcal{R}_1 = \mathcal{R}_0\left(1 + \frac{\imath_1}{\mathcal{R}_0}\right) = \mathcal{R}_0(1 + \imath_{n1}) = \mathcal{R}_0\left(1 - \frac{y'}{g_0}\right) = \mathcal{R}_0(1 - y'_n),$$

$$\mathcal{R}_2 = \mathcal{R}_0\left(1 + \frac{\imath_2}{\mathcal{R}_0}\right) = \mathcal{R}_0(1 + \imath_{n2}) = \mathcal{R}_0\left(1 + \frac{x'}{g_0}\right) = \mathcal{R}_0(1 + x'_n),$$

$$\mathscr{R}_3 = \mathscr{R}_0\left(1 + \frac{\imath_3}{\mathscr{R}_0}\right) = \mathscr{R}_0(1 + \imath_{n3}) = \mathscr{R}_0\left(1 + \frac{y'}{g_0}\right) = \mathscr{R}_0(1 + y'_n),$$

and

$$\mathscr{R}_4 = \mathscr{R}_0\left(1 + \frac{\imath_4}{\mathscr{R}_0}\right) = \mathscr{R}_0(1 + \imath_{n4}) = \mathscr{R}_0\left(1 - \frac{x'}{g_0}\right) = \mathscr{R}_0(1 - x'_n),$$

in which

$$\mathscr{R}_0 = \frac{g_0}{\mu A},$$

$$\Phi_1 = \Phi_0\left(1 + \frac{\phi_1}{\Phi_0}\right) = \Phi_0(1 + \phi_{n1}),$$

$$\Phi_2 = \Phi_0\left(1 + \frac{\phi_2}{\Phi_0}\right) = \Phi_0(1 + \phi_{n2}),$$

$$\Phi_3 = \Phi_0\left(1 + \frac{\phi_3}{\Phi_0}\right) = \Phi_0(1 + \phi_{n3}),$$

and

$$\Phi_4 = \Phi_0\left(1 + \frac{\phi_4}{\Phi_0}\right) = \Phi_0(1 + \phi_{n4}).$$

In these equations the currents and fluxes are regarded as rms complex quantities. Use of these normalized quantities and Q_0 and Q in Equations (7-1) through (7-11) gives

$$i_{n1} + i_{n2} = \imath_{n1} + (1 + \imath_{n1})\phi_{n1} + \imath_{n2} + (1 + \imath_{n2})\phi_{n2}, \tag{7-12}$$

$$i_{n2} + i_{n3} = \imath_{n2} + (1 + \imath_{n2})\phi_{n2} + \imath_{n3} + (1 + \imath_{n3})\phi_{n3}, \tag{7-13}$$

$$i_{n3} + i_{n4} = \imath_{n3} + (1 + \imath_{n3})\phi_{n3} + \imath_{n4} + (1 + \imath_{n4})\phi_{n4}, \tag{7-14}$$

$$i_{n4} + i_{n1} = \imath_{n4} + (1 + \imath_{n4})\phi_{n4} + \imath_{n1} + (1 + \imath_{n1})\phi_{n1}, \tag{7-15}$$

and

$$\phi_{n1} + \phi_{n3} = \phi_{n2} + \phi_{n4}, \tag{7-16}$$

only three of Equations (7-12) through (7-15) being independent;

$$i_{n1} = \frac{Q_0 \phi_{n1}}{1 - j(Q_0 - Q)}, \tag{7-17}$$

$$i_{n2} = \frac{Q_0 \phi_{n2}}{1 - j(Q_0 - Q)}, \tag{7-18}$$

$$i_{n3} = \frac{Q_0 \phi_{n3}}{1 - j(Q_0 - Q)}, \tag{7-19}$$

and

$$i_{n4} = \frac{Q_0 \phi_{n4}}{1 - j(Q_0 - Q)}; \tag{7-20}$$

$$F_{x'} = \frac{\Phi_0^2}{2\mu A}[(1 + \phi_{n4})^2 - (1 + \phi_{n2})^2] = \frac{N^2 V^2 \mu A}{2R^2 g_0^2} \frac{(1 + \phi_{n4})^2 - (1 + \phi_{n2})^2}{Q^2 + 1}$$

$$= \frac{N^2 V^2 \mu A}{2R^2 g_0^2} F_{x'n}, \tag{7-21}$$

$$F_{y'} = \frac{\Phi_0^2}{2\mu A}[(1 + \phi_{n1})^2 - (1 + \phi_{n3})^2] = \frac{N^2 V^2 \mu A}{2R^2 g_0^2} \frac{(1 + \phi_{n1})^2 - (1 + \phi_{n3})^2}{Q^2 + 1}$$

$$= \frac{N^2 V^2 \mu A}{2R^2 g_0^2} F_{y'n}, \tag{7-22}$$

and

$$F_r = \sqrt{F_{x'}^2 + F_{y'}^2}. \tag{7-23}$$

In Equations (7-21) and (7-22) the additions in $(1 + \phi_n)^2$ are complex, but the subtractions of the squared terms are with respect to magnitude only. The simultaneous solutions of Equations (7-12) through (7-20) for selected values of normalized displacements x'_n and y'_n (which define the i's), Q_0, and Q give the corresponding normalized current and flux departures and the normalized force components. For the force components the normalizing factor is the pull of *one pole* for resonant current with the cylinder centered, whereas for the single-axis suspension the normalizing factor is the pull on one end of the block, or two poles, for resonant current with the block centered.

The stiffness along either axis with the cylinder centered can be derived[43,44] by computing the restoring force along one axis (with the coils for the other axis normally connected), dividing the expression by the displacement, and then allowing the displacement to approach zero, just as is done for the single-axis suspension. The result is

$$F_{x'}\bigg|_{x'\to 0} = F_{y'}\bigg|_{y'\to 0} = \frac{N^2V^2\mu A}{2R^2g_0^3}\frac{4(Q^2 - Q_0Q + 1)}{(Q^2 + 1)^2} = \frac{N^2V^2\mu A}{2R^2g_0^3}F_{n0}, \qquad (7\text{-}24)$$

and in fact this stiffness applies in all directions, radially.

Position signals can be obtained for displacements along either axis just as for the single-axis suspension by utilizing bridge circuits formed from circuits 4 and 2 for the x' axis and from circuits 1 and 3 for the y' axis. Also, the quiescent position can be adjusted by the use of trimming capacitance.

7-3 Derivation of Equations for the Eight-Pole Cylindrical Magnetic Suspension[6,7,8,11,14,15]

An approximate solution for the behavior of this suspension can be obtained by imagining the stator structure to be broken as shown in Figure 7-5, so that each coil pair has an independent magnetic circuit. The equations relating to displacement along either axis then are the same as already derived for the single-axis suspension except for the projecting of displacements and forces along the respective axes. Such adjustment of Equation (2-7) gives

$$F_x = F_y = \frac{N^2V^2\mu A}{4R^2g_0^2}\frac{4(Q^2 - Q_0Q + 1)x_n\cos^2\gamma}{\{[Q_0 - (Q_0 - Q)(1 - x_n\cos\gamma)]^2 + (1 - x_n\cos\gamma)^2\}} \\ \times \{[Q_0 - (Q_0 - Q)(1 + x_n\cos\gamma)]^2 + (1 + x_n\cos\gamma)^2\}}.$$

$$(7\text{-}25)$$

This equation is based on the assumption that when displacement is along one axis the poles that straddle the other axis produce no net force along the direction of displacement; this assumption is not exactly correct but is very nearly correct for very small displacements. Therefore the stiffness along either axis when the cylinder is centered, as derived from Equation (7-25),

$$F_x = F_y = \frac{N^2V^2\mu A}{4R^2g_0^3}\frac{4(Q^2 - Q_0Q + 1)\cos^2\gamma}{(Q^2 + 1)^2} = \frac{N^2V^2\mu A}{4R^2g_0^3}F_{n0}, \qquad (7\text{-}26)$$

is essentially correct for the eight-pole structure as built, as shown in Figure 7-2, because when the cylinder is centered the circuits in effect are not coupled. However, Equation (7-26) is not quite correct for stiffness in other radial directions. For small displacements, Equation (7-25) is a good approximation

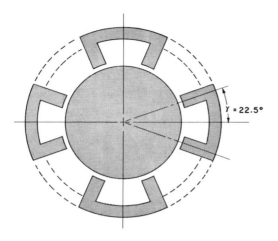

Figure 7-5 Structure for eight-pole magnetic suspension with independent magnetic circuits.

for restoring-force components if applied to the structure of Figure 7-2, but for large displacements it can be considerably in error. For simplicity and strength a stator structure represented by Figure 7-2 commonly is used, so that a rigorous analysis requires a procedure similar to the one used for the four-pole structure, owing to the multiplicity of couplings.

The following example also is written for the series-tuned connection with voltage source, Case (1), but the equations are placed immediately in normalized form. The magnetic-circuit equations become

$$2i_{n81} = i_{n8} + (1 + i_{n8})\phi_{n8} + i_{n1} + (1 + i_{n1})\phi_{n1}, \tag{7-27}$$

$$-i_{n81} + i_{n23} = -i_{n1} - (1 + i_{n1})\phi_{n1} + i_{n2} + (1 + i_{n2})\phi_{n2}, \tag{7-28}$$

$$2i_{n23} = i_{n2} + (1 + i_{n2})\phi_{n2} + i_{n3} + (1 + i_{n3})\phi_{n3}, \tag{7-29}$$

$$i_{n23} - i_{n45} = i_{n3} + (1 + i_{n3})\phi_{n3} - i_{n4} - (1 + i_{n4})\phi_{n4}, \tag{7-30}$$

$$2i_{n45} = i_{n4} + (1 + i_{n4})\phi_{n4} + i_{n5} + (1 + i_{n5})\phi_{n5}, \tag{7-31}$$

$$-i_{n45} + i_{n67} = -i_{n5} - (1 + i_{n5})\phi_{n5} + i_{n6} + (1 + i_{n6})\phi_{n6}, \tag{7-32}$$

$$2i_{n67} = i_{n6} + (1 + i_{n6})\phi_{n6} + i_{n7} + (1 + i_{n7})\phi_{n7}, \tag{7-33}$$

$$i_{n67} - i_{n81} = i_{n7} + (1 + i_{n7})\phi_{n7} - i_{n8} - (1 + i_{n8})\phi_{n8}, \tag{7-34}$$

and

$$\phi_{n1} + \phi_{n2} + \phi_{n5} + \phi_{n6} = \phi_{n3} + \phi_{n4} + \phi_{n7} + \phi_{n8}. \tag{7-35}$$

Of Equations (7-27) through (7-34) only seven are independent. In those equations,

$$\imath_{n1} = -y_n \sin \gamma - x_n \cos \gamma,$$

$$\imath_{n2} = -y_n \cos \gamma - x_n \sin \gamma,$$

$$\imath_{n3} = -y_n \cos \gamma + x_n \sin \gamma,$$

$$\imath_{n4} = -y_n \sin \gamma + x_n \cos \gamma,$$

$$\imath_{n5} = y_n \sin \gamma + x_n \cos \gamma,$$

$$\imath_{n6} = y_n \cos \gamma + x_n \sin \gamma,$$

$$\imath_{n7} = y_n \cos \gamma - x_n \sin \gamma,$$

and

$$\imath_{n8} = y_n \sin \gamma - x_n \cos \gamma.$$

The electric-circuit equations are

$$i_{n81} = -\frac{jQ_0}{1 - j(Q_0 - Q)}\left(\frac{\phi_{n8} + \phi_{n1}}{2}\right), \tag{7-36}$$

$$i_{n23} = -\frac{jQ_0}{1 - j(Q_0 - Q)}\left(\frac{\phi_{n2} + \phi_{n3}}{2}\right), \tag{7-37}$$

$$i_{n45} = -\frac{jQ_0}{1 - j(Q_0 - Q)}\left(\frac{\phi_{n4} + \phi_{n5}}{2}\right), \tag{7-38}$$

and

$$i_{n67} = -\frac{jQ_0}{1 - j(Q_0 - Q)}\left(\frac{\phi_{n6} + \phi_{n7}}{2}\right). \tag{7-39}$$

The force components are

$$F_x = \frac{N^2 V^2 \mu A}{8R^2 g_0^2}\left[\frac{(1 + \phi_{n8})^2 + (1 + \phi_{n1})^2 - (1 + \phi_{n4})^2 - (1 + \phi_{n5})^2}{Q^2 + 1}\cos \gamma \right.$$

$$\left. + \frac{(1 + \phi_{n2})^2 + (1 + \phi_{n7})^2 - (1 + \phi_{n3})^2 - (1 + \phi_{n6})^2}{Q^2 + 1}\sin \gamma \right],$$

$$\tag{7-40}$$

$$F_y = \frac{N^2 V^2 \mu A}{8R^2 g_0^2} \left[\frac{(1 + \phi_{n2})^2 + (1 + \phi_{n3})^2 - (1 + \phi_{n6})^2 - (1 + \phi_{n7})^2}{Q^2 + 1} \cos \gamma \right.$$

$$\left. + \frac{(1 + \phi_{n1})^2 + (1 + \phi_{n4})^2 - (1 + \phi_{n5})^2 - (1 + \phi_{n8})^2}{Q^2 + 1} \sin \gamma \right],$$

$$(7\text{-}41)$$

and

$$F_r = \sqrt{F_x^2 + F_y^2}. \tag{7-42}$$

The simultaneous solution of Equations (7-27) through (7-39) for selected values of normalized displacements x_n and y_n (which together with $\gamma = 22.5°$ determine the ι_n's), Q_0, and Q give the corresponding normalized current and flux departures and the normalized force components.

Position signals can be obtained for displacement along either axis by utilizing bridge circuits formed from circuits 8-1 and 4-5 for the x axis and from circuits 2-3 and 6-7 for the y axis. Also the quiescent position can be adjusted by use of trimming capacitance.

Equations for the eight-pole magnetic suspension combined with the Microsyn can be written in exactly the same manner as the equations for the eight-pole magnetic cylindrical suspension, except that the reluctance departures then become functions of small polar area changes as well as functions of the displacements. For comparable frame sizes the combined device, Figure 7-3, is inherently the weaker because of the larger projection angles, the smaller gap areas, and the smaller space available for the suspension windings when two sets of windings are required.

7-4 Comparison of Structures; Effects of Magnetic Couplings[15]

When displacement occurs along one axis, the poles that are along or straddle the other axis exert very little force along the displacement direction. For the four-pole cylindrical suspension, the change in reluctance of the y'-axis gaps is very small for small displacement of the cylinder in the x' direction, and vice versa. For the eight-pole cylindrical suspension, the change in total reluctance of a pair of gaps for poles that straddle the y axis is very small for a small displacement of the cylinder in the x direction, and vice versa. For the eight-pole combined magnetic suspension and Microsyn, the corresponding reluctance changes are somewhat larger, as can be appreciated by observation of Figure 7-3, owing to the overlap of rotor and stator poles. However, apart from these small reluctance forces, the primary reason for influence of restoring force along one axis owing to the excitation of coils on poles that are along or straddle an orthogonal axis is the change in magnetic couplings among the windings with displacement of the cylinder. As an illustration, Figure 7-6 shows the measured force-displacement relations for an eight-pole cylindrical magnetic suspension operated with voltage source and series tuning, first with all coils excited and then with only the coils that straddle the displacement axis excited. The difference is not because of any significant change in force components along the displacement axis at the poles that straddle the other axis but because the currents in the coils on those poles substantially change the fluxes in the gaps under the other poles. The curve for excitation of only the coils that straddle the displacement axis is as expected for a single-axis suspension and approaches zero force asymptotically as displacement increases, whereas with all coils excited the force magnitude is somewhat smaller and eventually reverses sign.

In Figure 7-7 comparative force-displacement curves are plotted for all the various structures considered, for displacement along one axis. These curves are for series tuning with $Q = 1$ and use of a voltage source, the same total polar cross-sectional area,* and the same flux densities in the gaps when the suspended members are centered. Other bases of comparison of course would give different results. As previously indicated, the combined device, Figure 7-3, is largely out of use. In Table 7-1 comparative stiffnesses of the four-pole and eight-pole cylindrical suspensions are shown as functions of directions of radial displacement. This comparison also is for series tuning and use of a voltage source with $Q = 1$ and the same flux densities in the gaps when the cylinders are centered. These numbers are obtained from computer solutions and are not stiffnesses at zero displacement but are averages over small displacements r_n. In this comparison one must remember that for $x_n = x'_n$ the displacement along a polar axis is only 0.924 as much for the eight-pole as for the four-pole cylindrical suspension.

7-5 Circuit Connections for Four-Pole and Eight-Pole Magnetic Suspensions[45]

All circuit connections illustrated for the single-axis suspensions are applicable to the two-axis suspensions, one pair of circuits or one bridge circuit applying to each axis, except that for the four-pole suspension special winding arrangements are necessary to use the bridge circuit because four coils cannot

---

*For the Microsyn the polar overlap is taken as half the polar area, but the fraction need not be one-half.

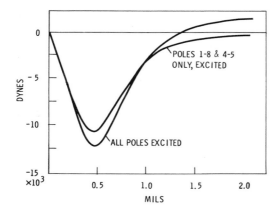

Figure 7-6 Influence of magnetic couplings on centering force, eight-pole magnetic suspension.

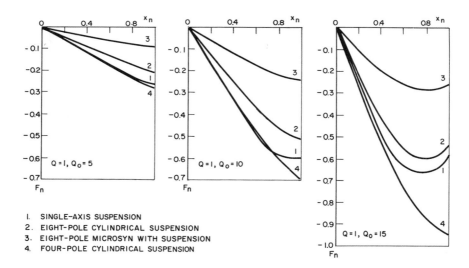

Figure 7-7 Comparison of normalized centering force versus normalized displacement for various magnetic suspensions.

Table 7-1. Directional Variation of Normalized Stiffness

Directional Deflection				F_n/r_n			
x_n or x_n'	y_n or y_n'	r_n	Angle	$Q_0 = 5$	$Q_0 = 10$	$Q_0 = 15$	$Q_0 = 20$
Four-pole Cylindrical Suspension							
0.0400	0.0000	0.0400	0.0°	2.98	7.90	12.65	17.10
0.0400	0.0200	0.0447	26.5°	2.98	7.85	12.55	16.80
0.0400	0.0400	0.0567	45.0°	2.96	7.75	12.15	15.70
Eight-pole Cylindrical Suspension							
0.0400	0.0000	0.0400	0.0°	2.24	6.35	10.20	13.55
0.0400	0.0200	0.0447	26.5°	2.24	6.40	10.30	13.80
0.0400	0.0400	0.0567	45.0°	2.24	6.35	10.20	13.50

form two bridge circuits. However, if the circuit pairs are placed in parallel across a current source or if the circuit pairs are placed in series across a voltage source, displacement along one axis affects sensitivity along the other axis. The same situation exists for the parallel or series connections of bridge circuits. Various alternative connections are shown in Figures 7-8 through 7-12, the selection of which depends largely on the number of leads that must be brought out, the number of capacitors required, the type of source to be used, and the voltage or current capacity of the available supply. Equations for any of these circuits can be written for computer solution by methods already illustrated.

Two connections are shown in Figure 7-8 for Cases (1) and (2) that utilize series tuning. If a voltage source is used in Figure 7-8(a) or if a current source is used in Figure 7-8(b), displacement along one axis does not greatly affect sensitivity along the other axis; whereas if a current source is used in Figure 7-8(a) or if a voltage source is used in Figure 7-8(b), considerable such interfering interaction exists. The circuit of Figure 7-8(a) requires that five leads be brought out, whereas the circuit of Figure 7-8(b) requires that only four leads be brought out. The relative source voltage or current requirements for comparable performances are indicated on the figure.

Two connections are shown in Figure 7-9 for Cases (3) and (4) that utilize parallel tuning. If a voltage source is used in Figure 7-9(a) or if a current source is used in Figure 7-9(b), displacement along one axis does not greatly affect sensitivity along the other axis; whereas if a current source is used in

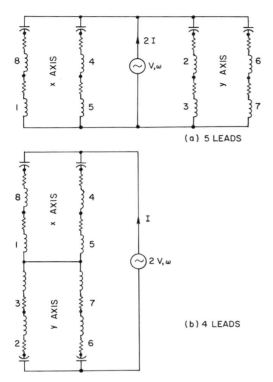

Figure 7-8 Circuits for two-axis adaptations of Case (1) or (2): (a) x- and y-axis circuits in parallel, (b) in series.

Figure 7-9(a) or if a voltage source is used in Figure 7-9(b), considerable such interfering action exists. The circuit of Figure 7-9(a) requires that only four leads be brought out, whereas the circuit of Figure 7-9(b) requires that five leads be brought out. The relative source voltage or current requirements for comparable performances are indicated on the figure.

Connections are shown in Figure 7-10 for Cases (5) through (8) that utilize bridge circuits.[39] These circuits require two, three, or four capacitors. If a voltage source is used as in Figure 7-10(a), with or without series tuning capacitors, the two bridge circuits act independently. If the series tuning capacitors are combined, as in Figure 7-10(c), one capacitor is eliminated, but then displacement along one axis affects sensitivity along the other axis. If a current source is used as in Figure 7-10(a), with or without a shunt tuning capacitor, displacement along one axis affects the sensitivity along the other

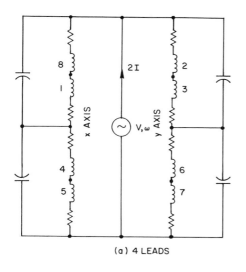

(a) 4 LEADS

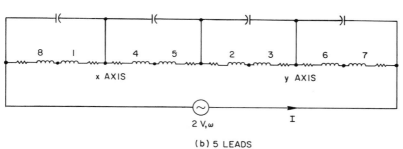

(b) 5 LEADS

Figure 7-9 Circuits for two-axis adaptations of Case (3) or (4): (a) x- and y-axis circuits in parallel, (b) in series.

axis. If a voltage source is used as in Figure 7-10(b), with or without a series tuning capacitor, displacement along one axis affects the sensitivity along the other axis. If a current source is used as in Figure 7-10(b) without a shunt tuning capacitor, the two bridge circuits act independently; but if a shunt tuning capacitor is used, displacement along one axis affects sensitivity along the other axis.

In all of these circuits except the bridge circuits, the coil pairs of the eight-pole suspensions can be replaced by the corresponding single coils of the four-pole suspension. To use the bridge-circuit connections with the four-coil suspension merely requires that two equal coils be placed on each pole;

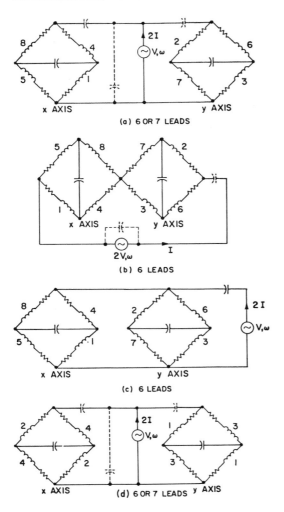

Figure 7-10 Circuits for two-axis adaptations of Case (5), (6), (7), or (8): (a) x- and y-axis circuits in parallel, (b) in series, (c) for Case (7) with tuning capacitors combined, x- and y-axis circuits in parallel, (d) bridge-circuit arrangement for four-pole suspension.

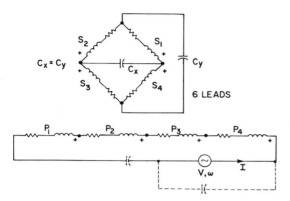

Figure 7-11 Special apparent bridge-circuit arrangement for four-pole two-axis suspension.

one set can form one bridge circuit, and the other set can form another bridge circuit, as shown in Figure 7-10(d), as an example. To make the coils of each pair equal, they can be wound simultaneously as a bifilar winding. A special arrangement shown in Figure 7-11 can adapt the four-pole suspension to an apparent "bridge" circuit through transformer action, wherein the primaries are connected in series aiding, in accorance with the flux directions of Figure 7-1, across a voltage or a current source; and the secondries are connected in series in a closed loop with alternate polarities, so that when the cylinder is centered the secondaries carry no current. When the cylinder is displaced, the capacitances connected across alternate secondary terminals cause the net magnetomotive forces on poles that face the shorter gaps to decrease and the net magnetomotive forces on poles that face the longer gaps to increase suffi-ciently to produce a centering action. The scheme requires only two capaci-tors, or three if tuned, and six leads.

A somewhat unique arrangement is illustrated in Figure 7-12(a). Whereas this circuit might qualify under the general definition of a bridge circuit, it is called a *mesh* circuit[46] to distinguish it from the simple bridge circuits already discussed. It really is an outgrowth of the original active circuitry[1,2*] in which the alternating current was used to provide feedback position signals. A feature of this mesh circuit, in contrast with the bridge circuit, is that only one mesh is required for two-axis control, whereas two bridge circuits are required for such control. Further, this mesh circuit requires only four leads

---

* Further discussion of this mesh circuit is in Section 9-1.

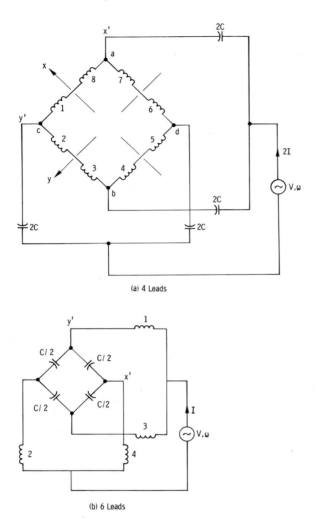

(a) 4 Leads

(b) 6 Leads

Figure 7-12 Mesh circuit for two-axis suspension: (a) square of coils, (b) square of capacitors.

to be brought out, in contrast with five to seven leads for some two-axis circuits. Also the mesh circuit is equally applicable to the four-pole or the eight-pole suspension. In Figure 7-12(a), eight coils are shown, but the paired coils may be taken to represent the individual coils of a four-pole suspension if the primed and unprimed axes are rotated 45° clockwise. The voltage across points a-b gives the displacement signal for the x' axis, and the voltage across points c-d gives the displacement signal for the y' axis, for the eight-pole suspension.

A disadvantage of this mesh circuit is that it requires twice the capacitance of other two-axis circuits shown. With the same source voltage as the circuit of Figure 7-8, for example, which is the two-axis version of Case (1), series tuning, with a voltage source, the initial stiffness of this mesh circuit is only about 0.033 as much, for $Q = 1$ and $Q_L = 10$, and is somewhat anomalous in that the initial stiffness decreases very considerably with increase in Q_L. For equal current sources $2I$ (which give the same quiescent coil currents) the mesh circuit of Figure 7-12(a) gives half the initial stiffness of the series tuning connection of Figure 7-8.

A modification of the mesh circuit is shown in Figure 7-12(b), in which the capacitors rather than the coils form a square. This circuit also requires only one mesh for two-axis control but requires six leads to be brought out. It also is equally applicable to the four-pole or the eight-pole suspension; the four-pole version is shown. It requires only half the capacitance of the other two-axis circuits shown. With the same source voltage as the circuit of Figure 7-8, it develops only about 0.018 as much stiffness for $Q = 1$ and $Q_L = 10$, and the stiffness decreases even more with increase of Q_L than for the connection of Figure 7-12(a). For current source I (which gives quiescent coil currents the same as for the Figure 7-8 connection) the mesh circuit of Figure 7-12(b) gives twice the initial stiffness of the series-tuning connection of Figure 7-8.

7-6 Three-Axis Combination of Suspensions[8,11,14,15]

By making the gap surfaces of a two-axis suspension conical and using two such suspensions on the same axis, Figure 7-13, the combination can produce axial centering. The axial restoring forces are obtained at the expense of some weakening of the radial centering forces, and the relative axial and radial sensitivities depend on the angle α of the cone. Further, the radial sensitivities are affected by axial displacements, owing to change of gap lengths, and axial forces may be produced by radial displacements. Radial restoring forces are

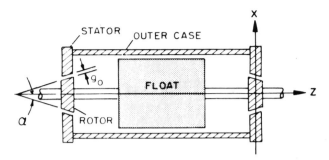

Figure 7-13 Three-axis suspension using two units having conical gap surfaces.

the net differences of forces normal to the gaps, projected normal to the common axis of the suspensions; whereas axial forces are the net sums of forces normal to the gaps, projected along the common axis of the suspensions. Hence if the two suspensions are subjected to different radial displacements and no axial displacement whatever by inertia forces, a net axial force can develop and cause a temporary axial displacement that must correct itself. Not all of the various ways of connecting magnetic suspension circuits that have been considered can be used to give restoring forces axially merely through use of conical gap faces, as is explained presently.

In deriving a relation between radial and axial stiffness of an eight-pole magnetic suspension at zero displacement, the small variations in radial stiffness with radial direction is ignored. If g is the displacement normal to a gap surface, r is the corresponding radial displacement, and z is the corresponding axial displacement, then

$$r = \frac{g}{\cos \gamma \cos \alpha} \tag{7-43}$$

and

$$z = \frac{g}{\sin \alpha}. \tag{7-44}$$

For zero radial and axial displacements, the radial stiffness is

$$\dot{F}_r \bigg|_{r \to 0} = \frac{F_g \cos \gamma \cos \alpha}{r} \bigg|_{r \to 0} = \frac{F_g \cos^2 \gamma \cos^2 \alpha}{g} \bigg|_{g \to 0} = \frac{F_g \cos^2 \gamma \cos^2 \alpha}{g_n g_0} \bigg|_{g_n \to 0}, \tag{7-45}$$

and the axial stiffness is

$$\dot{F}_z\bigg|_{z\to 0} = \frac{4F_g \sin\alpha}{z}\bigg|_{z\to 0} = \frac{4F_g \sin^2\alpha}{g}\bigg|_{g\to 0} = \frac{4F_g \sin^2\alpha}{g_n g_0}\bigg|_{g_n\to 0}, \tag{7-46}$$

in which

$$g_n = \frac{g}{g_0}$$

and

$$F_g = F_x = F_y$$

for the corresponding cylindrical suspension. Hence for the combined conical suspensions, the ratio of axial to radial stiffness with zero radial and axial displacements is

$$\frac{\dot{F}_{z0}}{\dot{F}_{r0}} = \frac{\dot{F}_z|_{z\to 0}}{\dot{F}_r|_{r\to 0}} = \frac{4\sin^2\alpha}{\cos^2\gamma\cos^2\alpha} = \frac{4\tan^2\alpha}{\cos^2\gamma}. \tag{7-47}$$

For the eight-pole cylindrical suspension, $\gamma = 22.5°$, so that for equal axial and radial stiffnesses of the combined conical suspensions at zero displacement

$$4\tan^2\alpha = (0.924)^2, \quad \tan\alpha = 0.462, \quad \alpha = 24.8°.$$

For the four-pole suspension, the derivation for the eight-pole suspension may be used by taking $\gamma = 0$, so that for equal axial and radial stiffnesses of the combined conical suspensions at zero displacement

$$4\tan^2\alpha = 1, \quad \tan\alpha = 0.500, \quad \alpha = 26.6°.$$

Other conical slopes α may be desirable. For example, the damping action of the flotation fluid for a suspended instrument ordinarily is such that axial displacement is much more sluggish than radial displacement in response to inertia forces. Therefore, to keep the device centered requires much less axial stiffness than radial stiffness. On the other hand to center the device originally, or erect it, with a relatively weak axial force could require a very long time. Hence the determination of the ratio of axial to radial stiffness, or the slope of the cone, is one of the designer's problems depending on the overriding requirements of the application.

For suspensions having more than eight poles, the ratio of axial to radial stiffness is a more complicated relation, because poles or pole pairs not centered on the x or y axis contribute to the pull.

Changes in radial force-displacement characteristics caused by axial displacement are illustrated in Figure 7-14 for an eight-pole suspension connected as indicated. For each axially displaced position, the suspension acts radially as if the quiescent gap g_0 had been changed, with accompanying changes in Q_0 and Q. An idea of the action of radial displacements in producing axial forces may be obtained by plotting the sums of the opposing pulls at the two ends of the single-axis suspension. Such plots are given in normalized form in Figure 7-15 for Case (1) or (7), with flux densities adjusted to the 0.735 normalized limit. These curves can be translated to the fixed voltage-source condition by dividing the ordinates by the squares of the appropriate translation factors of Table 5-3. Thus the force sums would be increased by factors of 1.21 to 1.69, depending on the value of Q, for that condition. For a radial displacement along the x axis at one end of the z axis, the displacement normal to the tapered gap surfaces of the poles that straddle the x axis at that end of

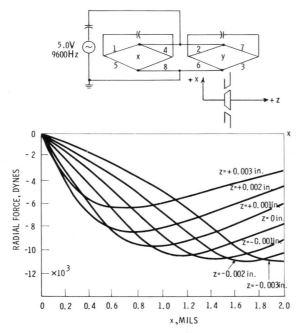

Figure 7-14 Changes in radial force-displacement characteristic caused by axial displacement, Case (7) connection, Figure 7–10(a).

the z axis is

$g_1 = x_1 \cos \gamma \cos \alpha;$

and if Figure 7-15 is entered with the normalized displacement

$$g_{n1} = \frac{g_1}{g_0}$$

and the corresponding normalized force sum F_{n1} is obtained, then the axial force at end 1 for a structure such as Figure 7-5 but with tapered gaps is

$F_{n1a} = F_{n1} \cos \gamma \sin \alpha.$

Similarly, for a radial displacement along the x axis at the other end of the z axis,

$F_{n2a} = F_{n2} \cos \gamma \sin \alpha,$

and the net force caused by the x-axis displacements is

$F_{na} = F_{n1a} - F_{n2a} = (F_{n1} - F_{n2}) \cos \gamma \sin \alpha.$

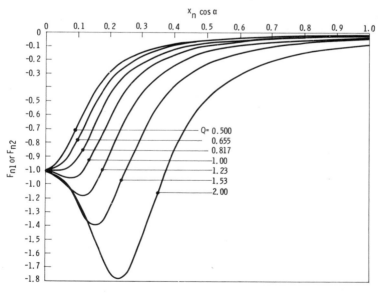

Figure 7-15 Normalized axial force versus normalized radial displacement, Case (1) or (7), limited maximum flux density.

Design should be such that for maximum allowable axial displacement one or the other suspension does not become unstable radially and for maximum allowable radial displacements intolerable axial forces do not result. This situation can be a problem in erection as well as during operation and can create a situation sometimes called "axial-radial hangup."

The circuit of Figure 7-8 or 7-10 made with a voltage source and series tuning could be advantageous in this situation, because when displacement occurs along one axis the part of the circuit that essentially controls the performance along that axis receives the larger share of the applied voltage. For displacement in a direction midway between the two axes, the applied voltage would divide equally between the two parts of the circuit, so that the performance would be the same as for Figure 7-8(a) or 7-10(a) made with a voltage source and series tuning. As the direction of the displacing force approaches one axis, the component of stiffness along that axis increases and the component of stiffness along the other axis decreases, so that the increased radial restoring force is most pronounced for displacements directed in the vicinity of one axis or the other. For example, for $Q = 1$, $Q_0 = 10$, $g_n = 0.10$, $y = 0$, and $z = 0$, the division of voltage between the parts of the circuit of Figure 7-8(b) that essentially control the response along the respective axes are

$$V_x \approx 1.08V$$

and

$$V_y \approx 0.96V.$$

If now with g unchanged, which means x unchanged also, a z-axis deflection is established to the extent that

$$g_0' = 1.05g_0,$$

then

$$g_n' = \frac{0.10}{1.05} = 0.095,$$

$$Q_0' \approx 9.5,$$

$$Q' \approx 0.5,$$

$$z = \frac{g}{\sin \alpha},$$

and for equal radial and axial sensitivities

$$z_n = \frac{z}{g_0} = \frac{g}{g_0 \sin \alpha} = \frac{g_n}{\sin 24.8°} = \frac{0.10}{0.420} = 0.238,$$

which is a considerable axial deflection. Then

$$V_x' \approx 1.15V$$

and

$$V_y' \approx 0.85V.$$

In the first example, $y = 0$, $z = 0$, $g_n = 0.10$, the forces normal to the gap faces, and hence the axial force, are $(V_x/V)^2 = 1.17$ times as large as they would be if the radial displacement were made midway between the x and y axes, or with connections as for Figure 7-8(a) or 7-10(a) made with a voltage source and series tuning. In the second example, $y = 0$, $z_n = 0.238$, $g_n' = 0.095$, the axial force is $(V_x'g_0/Vg_0')^2 = (1.15/1.05)^2 = 1.20$ times as much as it would be if the radial displacement were made midway between the x and y axes, or with connections as for Figure 7-8(a) or 7-10(a) made with a voltage source and series tuning. However, the force for displacement midway between the x and y axes is considerably larger for the first example, $Q_0 = 10$, $Q = 1$, than for the second example, $Q_0 = 9.5$, $Q = 0.5$, being about 1.5 times as large. Whereas these examples show relatively large forces and large displacements, conversely, for equal disturbances, the increased stiffnesses resulting from the connections of Figure 7-8(b) or Figure 7-10(b) tend to limit the resulting displacements.

The illustrations used thus far have utilized series tuning with a voltage source, adaptations of Case (1). The tuned bridge circuits, adaptations of Case (7), perform similarly. If a current source is used, adaptations of Case (2) or (6), additional complications arise from the fact that radial displacements at one end of the z axis influence the radial sensitivity at the other end of the z axis. Further, if parallel tuning or the untuned bridge circuit is used, the arrangement is unstable along the z axis unless a modification is made in the circuitry. One such modification is shown in Figure 7-16, which really is a bridge circuit of bridge circuits. As for all the magnetic-suspension bridge circuits, drift of capacitance does not cause drift in position of the suspended member; but radial displacements cause axial displacements, and if radial displacements exist, change in axial displacement modifies them.

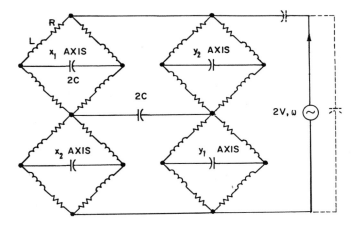

Figure 7-16 Bridge circuit of bridge circuits, to avoid z-axis instability of adaptations of Case (2) or (6).

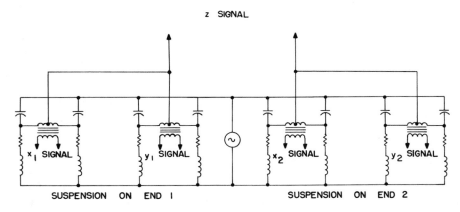

Figure 7-17 Circuitry for obtaining axial position signals, Case (1) adaptation, Figure 7-8(a).

Axial position signals can be obtained in the same manner as radial position signals by forming bridge circuits from corresponding circuits of the two suspensions that combine to give axial restraint, but when axial displacement is accompanied by radial displacement at one or both ends, the signals become somewhat garbled. One method of essentially eliminating the influence of radial displacement in obtaining an axial-displacement signal is illustrated in Figure 7-17 for the series-tuning case under discussion. Here the mid-taps of the transformers in the detector branches of the bridge circuits for the suspension at one end are essentially at the same potential, even if radial motion occurs, as is true also at the other end. However, if axial motion occurs, the potentials of the mid-taps at one end are different from the potentials of the mid-taps at the other end, and the difference of the potentials is indicative of axial displacement. The transformers being used with essentially open secondaries are assumed to present sufficiently high primary impedance to cause negligible disturbance of bridge-circuit conditions.

Quiescent axial position can be controlled by essentially equal trimming of the capacitances of the suspension circuits at one end with respect to the capacitances of the suspension circuits at the other end.

7-7 Other Three-Axis Arrangements

Axial centering can be achieved simply by allowing the pole pieces for the radial suspensions to overhang the ends of the suspended cylinder, or vice versa, as illustrated in Figure 7-18, parts (a) and (b); this polar arrangement makes use of the fringing fields at the ends of the structure and the fact that the magnetomotive force increases across the gap having the smaller permeance, owing to the tuning of the circuits. However, this method of axial constraint, though simple and requiring no extra circuitry, is rather weak.[4] Use of a rotor with I-shaped longitudinal cross section, Figure 7-18(c), can combine the radial stiffness of the Figure 1-5(a) structure and the longitudinal stiffness of the Figure 1-5(b) structure, but the rotor then cannot be advantageously laminated but should be a ferrite. However, this arrangement does not avoid interactions of radial and axial displacements. Independent control sometimes is desirable along the z axis, so that restraint along that axis may be made as stiff as is desired and interactions of radial and axial displacements are avoided. A variety of geometries may be used for this purpose, as illustrated by Figure 7-19, the parts and actions of which are largely self-explanatory and in which the circuitry has been omitted. Here only one end

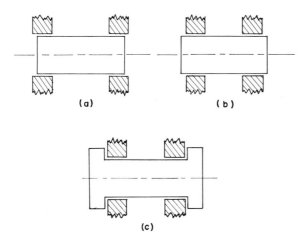

Figure 7-18 Axial centering: (a), (b) by use of fringing fields, (c) by use of rotor having "I" section.

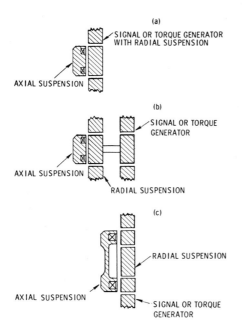

Figure 7-19 Separate radial and axial control: (a) radial suspension combined with Microsyn, (b) radial suspension separate from Microsyn, (c) Ducosyn.

of the device is shown, the other end being the same but reversed right and left.

The parts of Figure 7-19 differ only to illustrate different combinations with the Microsyn, E-type, or Ducosyn signal or torque generator. In Figure 7-19(a), the radial suspension and the Microsyn are supposed to be combined as in Figure 7-3. In Figure 7-19(b), the radial suspension is separate from the Microsyn or E-type signal or torque generator and operates on a cylinder. In Figure 7-19(c), a Ducosyn, illustrated in Figure 7-4, is supposed to be used.

Another way of achieving suspension independently along three axes is to use U-magnets at the ends of three orthogonal axes of a sphere or spherical shell of magnetic material, Figure 7-20(a), magnets at opposite ends of the same axis being paired in action as in Figure 2-1. Whereas this arrangement provides suspension, the sphere is free to float in any orientation unless some additional control is provided, but the device can be used as a spherical accelerometer. Such a device has been made and is described in Chapter 11, but it used a magnetic stator structure similar to the stator of the four-pole cylindrical suspension, Figure 7-1, but adapted to three dimensions, as shown in Figure 7-20(b). The circuitry then can be similar to the circuitry of Figure 7-1, utilizing six circuits, which of course are magnetically coupled through the stator structure, so that when the sphere is displaced, the restoring forces are not independent of each other. For small displacements the magnetic coupling is small, and the static displacements are measures of the component accelerations along the respective axes. These displacements can be measured by means of bridge signals taken for each axis as shown in Figure 2-1 for the U-magnet arrangement.

7-8 Numbers of Poles Other than Four or Eight[15,45]

Theoretically two-axis cylindrical magnetic suspensions or their conical modifications for combination to give three-axis suspensions can be made with any even number of equal* poles beginning with four. To have performance with respect to x-axis and y-axis displacement identical, the number of such poles must be a multiple of four (4, 8, 12, 16, 20, ...) if the coil for each pole has its separate circuit; if the coils are paired to halve the numbers of tuning capacitors required, the number of such poles is limited to multiples

---

* Structures can be made having unequal polar areas, such as E-type structures with large central leg areas. While such geometry has been applied to signal and torque generators, it has not been applied to suspensions.

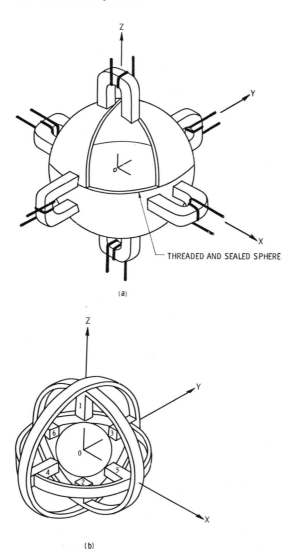

(a)

(b)

Figure 7-20 Three-axis spherical magnetic suspension: (a) using independent U magnets, Figure 2-1, (b) using adaptation of Figure 7-1 structure.

of eight (8, 16, 24, 32, …), that is, the even multiples of 4, the odd multiples of 4 giving different sensitivities with respect to the two axes. These same restrictions apply to the combined Microsyn and magnetic suspension, which is limited to use with a multiple of 4 poles in excess of 4, as has been stated previously. All other numbers of poles (6, 10, 12, 14, 18, 22, 26, …), whether the coil for each pole has its separate circuit or the coils are paired as previously indicated, give different sensitivities with respect to two orthogonal axes. In general, for a given number of equal poles, the use of a separate circuit for each coil gives less variations in sensitivity with respect to radial direction of displacement than the pairing of the coils to halve the number of circuits. When the coils have separate circuits, the sensitivity of a suspension is the same in a direction taken along the center line of any pole; when the coils are paired, the sensitivity of a suspension is the same along a line midway between the center lines of the poles carrying the paired coils. Within either arrangement, the variation of sensitivity with respect to radial direction decreases as the number of poles increases, but as a practical matter the number of poles is limited by the tolerable complexity of the circuitry and, in a given frame size, by space limitations and required strength of parts.

As an illustration, some normalized restoring forces and stiffnesses for a six-pole cylindrical magnetic suspension, Figure 7-21, are given in Table 7-2. Here the data are for series tuning with $Q = 1$, use of a voltage source, and the same total cross-sectional area and the same flux density in the gaps when the cylinder is centered as for the four- and eight-pole suspensions of Table 7-1. This type of six-pole suspension has been suggested for the possibility of reducing the number of circuits and tuning capacitors to three. Inspection of Table 7-2 shows not only that the sensitivity of this arrangement along the x axis is different from the sensitivity along the y axis but also that the sensitivity along the x axis depends on the direction of displacement along that axis. The dissymmetry of performance along the x axis is characteristic of any cylindrical magnetic suspension having an odd number of pole pairs with their coils paired. The dissymmetry in performance is due to the dissymmetry of pole distribution with respect to the y axis and is most pronounced for six poles, that is, for three pole pairs, and decreases as the number of poles increases to 10, 14, 18, 22, … . For the odd numbers of pole pairs, the magnetic polarity must alternate, or the dissymmetry is further aggravated.

Comparison with the data for the four- and eight-pole suspensions in Table 7-1 shows the six-pole suspension to be the weakest of the three. The reason can be appreciated by study of Figure 7-21 and observing the relatively small

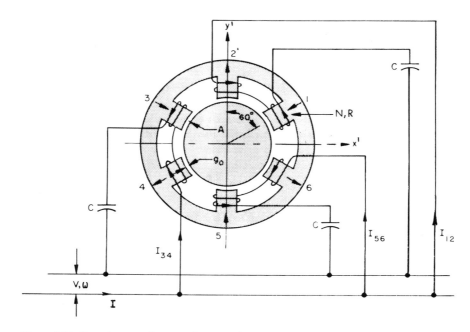

Figure 7-21 Six-pole two-axis magnetic suspension.

changes in some of the gap reluctances with respect to displacement along one axis or the other and the accompanying small contribution to force along one axis or the other, due to the large angles through which displacements and forces must be projected with respect to the axes. The equations for the six-pole suspension can be written in the same manner as illustrated for the four- and eight-pole suspensions for solution by computer, but since for displacement along the x axis, poles 2 and 5 contribute essentially no net force and since $\Phi_1 = \Phi_6$ and $\Phi_3 = \Phi_4$, a fairly simple solution can be made longhand for that case and is the source of data for Table 7-2:

$$\Phi_{n1} = \Phi_{n6} = \frac{[2 - j(Q_0 - 2)]x_n \sin 60^\circ}{2(1 - x_n \sin 60^\circ) + j[2(1 - x_n \sin 60^\circ) + Q_0 x_n \sin 60^\circ]} \quad (7\text{-}48)$$

and

$$\Phi_{n3} = \Phi_{n4} = -\frac{[1 - j(Q_0 - 1)]x_n \sin 60^\circ}{1 + x_n \sin 60^\circ + j[1 - (Q_0 - 1)x_n \sin 60^\circ]}, \quad (7\text{-}49)$$

Table 7-2. Normalized Restoring Forces and Stiffnesses for Six-Pole Suspension

x_n	0.00	+0.02	−0.02	+0.04	−0.04	+0.10	−0.10
y_n	±0.02	0.00	0.00	0.00	0.00	0.00	0.00
F_n	±0.102	−0.115	+0.113	−0.230	+0.188	−0.500	+0.430
$\dot{F}_n$	±5.10	−5.75	+5.65	−5.75	+4.70		

for $Q = 1$. For displacement along the y axis, longhand solution is more difficult, but for small displacements the accompanying changes in Φ_3 and Φ_4 can be neglected and an approximate solution obtained quite simply. Then $\Phi_1 = \Phi_2$ and $\Phi_5 = \Phi_6$, and poles 3 and 4 contribute no net force along the y axis:

$$\Phi_{n1} \approx \Phi_{n2} \approx \frac{[1 - j(Q_0 - 1)](1 + \cos 60°)y_n}{2 - (1 + \cos 60°)y_n + j[2 + (1 + \cos 60°)(Q_0 - 1)y_n]} \quad (7\text{-}50)$$

and

$$\Phi_{n5} \approx \Phi_{n6} \approx \frac{-[1 - j(Q_0 - 1)](1 + \cos 60°)y_n}{2 + (1 + \cos 60°)y_n + j[2 - (1 + \cos 60°)(Q_0 - 1)y_n]}. \quad (7\text{-}51)$$

These approximate equations were used to compute F_n for $y_n = \pm 0.02$ in Table 7-2, which is the reason why no larger values of y_n were used.

The suspensions that have different sensitivities along orthogonal axes, and in particular the ones that have dissymmetries, are not adaptable to giving position signals through bridge-circuit arrangements or readily through other special and more complicated arrangements, and in general they are not as practical as the four- and eight-pole arrangements.

7-9 Two-Axis Electric Suspensions[48,49]

Four- and eight-pole cylindrical electric suspensions analogous to the magnetic suspension can be developed, as indicated in Figure 7-22 and Figure 7-23, as in fact can the various multipole combinations mentioned for the magnetic suspension, though the only one tried in the Draper Laboratory has been the four-pole suspension with series tuning, using a voltage source. In general, the polar arcs for an electric suspension are made larger than the polar arcs of a magnetic suspension, because the electric suspension requires no interpolar space for coils. Therefore the polar force of a four-pole electric suspension may need to be integrated over the pole arc and resolved in the radial direction. As drawn, the eight-pole electric suspension of Figure 7-23

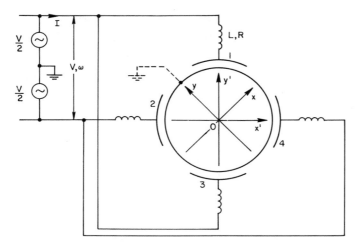

Figure 7-22 Four-plate electric suspension.

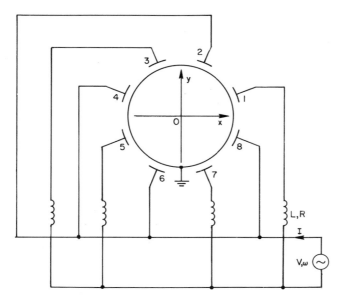

Figure 7-23 Eight-plate electric suspension with plates paired.

is analogous to the magnetic suspension having a frame as shown in Figure 7-2. If the rotor of Figure 7-23 were divided into four insulated segments, one opposite each pair of stator plates, and the ground removed, the arrangement would be analogous to a magnetic suspension having a frame as shown in Figure 7-5.

No composite electric suspension and signal generator or torquer analogous to the combined magnetic device of Figure 7-3 has been developed in the Draper Laboratory, but a combination somewhat similar to the Ducosyn, Figure 7-4, four-pole electric support inside, multipole electric signal generator or torquer outside, has been tried with limited success owing to electric shielding troubles and difficulties with the type of electric signal generator selected. In small devices, the volumes available for storage of electric-field energy with use of tolerable electric gradients permit the development of only rather small forces and torques compared with magnetic devices of about the same frame size.

Not all the circuit arrangements for the magnetic suspension, Figures 7-8 through 7-10, especially the bridge-circuit connections, can be translated to analogous circuits for the electric suspension. Unless the suspended cylinder is split into insulated segments, it is a common terminal for all the capacitors; this situation restricts the manner in which the circuits may be connected. With the electric suspension the problems of appropriate grounding and capacitance of leads and terminals may become acute, depending on the frequency used, the dimensions and geometric relations of the parts, and the environmental circumstances. The equations for the two-axis electric suspensions may be written for computer solution by methods analogous to the illustrations for the magnetic suspensions.

The extension of the electric suspension to three axes has not been tried in the Draper Laboratory but presumably could be accomplished either by use of conical geometry or by separate z-axis control.

7-10 Summary

All of the single-axis connections developed in Chapters 2, 3, and 4 can be used in various combinations of circuitry to give suspension along two orthogonal axes, and if the suspended member is a cylinder, the stiffness of the suspension can be approximately uniform radially. In general, the circuits are magnetically coupled by having their coils on the same structure, but such coupling can be essentially avoided by breaking the frame magnetically between adjacent coil pairs, Figure 7-5, but the fabrication of the device is thus complicated. In addition the circuits for x-axis and y-axis control may interfere with each other electrically, more or less, depending on the manner in which they are connected. Extension of control to three axes may be made simply by the use of conical gap faces, which permits significant interference between the axial and radial control circuits but requires no additional apparatus, or by utilizing separate z-axis control. Electric suspensions can be adapted to formation of two- and three-axis suspensions, but the variety of connections that can be used is limited by the fact that the suspended member may form a common connection for all the capacitors, and problems of isolation and grounding become difficult.

8 INFLUENCE OF MATERIALS

In the analysis thus far, aside from recognizing that coils have resistance, the structure and materials of the suspension have been regarded largely from the standpoint of their geometry in providing a framework and as media having ideal magnetic and dielectric properties. Actually, the physical properties of materials require modifications to be made in the analyses based on idealized materials. To some extent these modifications can be made with good enough approximation merely by adjustment of parameters used in the basic equations for the ideal device; sometimes the mathematical procedures need to be modified somewhat, and sometimes the effects are sufficiently drastic that mathematical analysis can serve only as a guide and principal reliance must be placed on experimentation.

8-1 Principal Considerations

Over the approximately 20-year period of development of the suspension, the frequency commonly used has advanced from 400 hertz to 12,800 hertz, with experimental applications as high as 40,000 hertz; also, square-wave and pulsed operation have come into use, as is discussed in the next chapter. The demand for high-precision performance has intensified. Hence the influence of such physical phenomena as electric and magnetic skin effects and eddy currents in general (with respect to energy losses, net magnetomotive force, and delay in flux change), magnetic hysteresis, magnetic saturation, magnetic disaccommodation, magnetic leakage fluxes, stray capacitances, dielectric losses, temperature changes and accompanying mechanical strains and changes in properties of materials, and mechanical misalignments and ellipticities due to imperfections of machine work and assembly procedures all increasingly challenge the skill of the designer in the determination of geometries and in the selection and processing of materials. Some of the most undesirable influences encountered when extreme precisions are demanded are second- or higher-order effects that are not fully predictable to the extent that they cannot be fully described or treated analytically or, if described adequately mathematically, can be handled only with extreme and unreasonable difficulty. In this area the designers' procedures, though relying on the guidance of science, emerge into the realm of engineering art as guided by intuition and experience.

8-2 Desired Properties of Magnetic Materials

In general the desirable properties of magnetic materials are high permeability, low core loss (both hysteresis and eddy-current losses), magnetic insensitivity to mechanical strain, and mechanical stability. These properties tend to keep the performance of the suspension largely independent of the materials of the magnetic structure. High permeability, low hysteresis and eddy-current effects, and low magnetostrictive effects tend to concentrate the magnetic effects into the air gaps. The heating due to small core losses is not in itself a serious influence, because the heat generated for ambient temperature control of the device is much in excess of the heat losses generated in the device. The principal effect of core losses may be to limit the achievable *effective* quality factor. Mechanical instabilities with respect to time or environmental conditions may affect the geometry of the device so as to introduce small torques or spurious position signals.

Commonly used magnetic materials are high-silicon iron alloys, such as the transformer steels; silicon-nickel-iron alloys, such as Sinimax; nickel-iron alloys, such as Monimax, Allegheny 4750, Armco 48, Carpenter 49, Hipernik, Hymu 80, or the various Permalloys or Mumetals; aluminum-iron alloys, such as Alfenol; and various soft ferrites, such as MN-31. High-silicon iron has the highest saturation flux density and the highest hysteresis loss but is comparable to the nickel-iron alloys in eddy-current loss and inferior in permeability. It is a relatively hard material. Sinimax is somewhat similar to Monimax, Allegheny 4750, Armco 48, Carpenter 49, and Hipernik, which are approximately 50 percent nickel materials, but has higher saturation flux density, lower eddy-current loss, and is somewhat harder. Hymu 80, the Permalloys, and the Mumetals, which are approximately 80 percent nickel materials, have the highest permeability and the lowest hysteresis loss of the group but a lower saturation flux density than the 50 Ni materials. The nickel-iron materials are relatively soft and ductile and are somewhat likely to distort in machining. Alfenol has a somewhat lower saturation flux density than the 50 Ni materials, a comparable permeability, and a lower eddy-current loss. It is relatively hard, machines well, and is mechanically more stable than the nickel-iron alloys. However, its magnetic properties are rather critical with respect to the percentage of aluminum and hence are uncertain of duplication from one melt to the next. Ferrites are the lowest in saturation flux density, permeability, and eddy-current loss but may have relatively high hysteresis loss. Their properties are relatively sensitive to temperature changes, and their Curie points are low. They are very stable mechanically but are very hard and brittle and are susceptible to cracking and chipping under stress. Finishing generally must be done by grinding. Being less dense than the ferromagnetic materials, ferrites can be used advantageously for reduction of mass and moment of inertia. All these materials are more or less strain sensitive and must be handled so as to minimize such effects. A summary of some representative properties for cold-rolled ferromagnetic sheets and a soft ferrite are given in Table 8-1:[50] saturation flux density $\mathscr{B}_s$, normal permeability μ at about 0.6 $\mathscr{B}_s$, coercive force $\mathscr{H}_c$ from saturation flux density, resistivity ρ, magnetostriction coefficient λ at about 0.6 $\mathscr{B}_s$, and machinability M. These data are merely nominal for comparative purposes. For magnetostriction effects the change in flux density or permeability as affected by stress is of more interest here than the elongation or shrinkage produced by the magnetic field, but the two effects are related.

Table 8-1. Representative Properties of Core Materials*

Core Material	$\mathscr{B}_s$ (gauss)	$\mathscr{H}_c$ (oersteds)	μ_r (relative)	ρ μohm-cm	λ in./in.	M
High Si-Fe	19,500	0.6	4,500	55	$+1 \times 10^{-6}$	good
Si-Ni-Fe	11,000	0.1	10,000	85		fair
50Ni-Fe	16,000	0.04	15,000	45	$+5 \times 10^{-6}$	fair
80Ni-Fe	7,500	0.02	20,000	55	0	fair
16Al-Fe	8,000	0.04	12,000	140	$+3 \times 10^{-6}$	good
Ferrite MN-31	3,500	0.1	4,000	25×10^8		poor

*These data are from Reference 50 and from manufacturers' literature.

Survey of this table reveals no one "best" material, so that the designer must make his compromises in accordance with his judgment in the light of the needs of the application and discussions with manufacturers and metallurgists.

Quite commonly *incremental* permeability as well as *normal* permeability has considerable influence in the design of active suspensions, as discussed in the next chapter. In such suspensions an alternating signal current may be superposed on a direct force current, or force-current pulses may be superposed on a quiescent current, either direct or alternating, so that the flux density resulting from the total excitation must be considered. For example, if an alternating-current signal is superposed on direct-current excitations, the signal sensitivity may be very low, owing to the low incremental permeability seen by the alternating current, even though the normal permeability may be high. This condition may become especially acute if the direct-current excitation places the flux density in the vicinity of saturation. Hence a high-saturation flux-density material can be advantageous in certain active suspensions.

8-3 Handling of Magnetic Materials

Much of the trouble that may arise from magnetic core materials, especially the more strain-sensitive materials, can be minimized by the appropriate selection of the material best suited to the application, by the proper handling, heat-treating, and mounting of the materials, and, if necessary, by the control of ambient temperature.

The properties of silicon-steel lamination material are well standardized. Manufacturers of high nickel-steel lamination material will certify the chemical analyses of the materials and supply the various physical properties and tolerances. A tight tolerance on stock thickness enables stacking and bonding with good parallelism. Depending on the geometry of the magnetic suspension, "transformer grade" or "rotor grade" should be specified. The former is oriented, whereas the latter is not oriented. Even with nonoriented laminations, the use of random stacking or the systematic rotation of each lamination with respect to its neighbor (considering the rolling direction of the sheet) may be advantageous in achieving a close approach to a homogeneous and isotropic structure with respect to the flux pattern. In punching the laminations, the punch and die should be kept sharp to ensure good flatness and to minimize burrs. The laminations should be thoroughly deburred either by tumbling or by stoning. Very thin laminations or laminations having complex geometry are not generally deburred successfully by tumbling because of the likelihood of becoming entangled and bent.

The manufacturer's recommended heat treatment for maximum permeability should be used. It should include the temperature cycle, the hydrogen dew point, the method of supporting the laminations, and the method of cleaning them, which is very important. Most material manufacturers either offer heat-treating service or will supply names and addresses of firms recommended for such work.

In bonding laminations, they should be stacked "make to break"; that is, since laminations are slightly cupped in the punching process, the cups all should face the same way. In bonding, a material should be used that does not introduce strains in the magnetic material upon hardening or creep as it ages. If the bonding agent does not provide extremely high resistance between laminations, an insulating surface should be created on each lamination. Ordinary oxidation generally is not satisfactory, because the oxide tends to stick to the bonding material better than it sticks to the metal, but special surfaces created by acid etching are very satisfactory. A tight tolerance on surface smoothness is helpful. Good insulation between laminations is important in minimizing eddy-current effects. A stack of laminations that shows high resistance between its outer surfaces is not necessarily satisfactory; the testing should be done by probing each pair of adjacent laminations. Unless all the interlamination resistance is good, eddy currents may exist in such a way as to distort the flux pattern considerably.

After heat treatment and bonding, the laminations should be subjected to an absolute minimum of machining, drilling, or grinding. Especially at the gap surfaces, if grinding is necessary, a plunge grind rather than a traverse grind should be used to avoid smearing the edges of laminations together and causing short circuits, and the grind should be as light as possible to avoid introduction of local strains and surface hardening that can distort the flux pattern in the gaps.

The manner of winding in place or the insertion, mounting, and connecting of coils may introduce strains in the magnetic materials, as may the manner of mounting the materials through tightening of bolts or clamps, forcing of shafts, pins, or jackets. Likewise the use of encapsulants requires great care and considerable know-how. Many potting compounds may introduce tensile, compressive, or twisting forces during the curing process that can rather drastically affect particularly the permeability and hysteresis of strain-sensitive magnetic materials. Different materials not only have different magnetostrictive sensitivities but behave oppositely. A material is said to have positive magnetostriction if its magnetization is increased by tension; the material also expands, and its hysteresis loop tends to be squared. The opposites happen under compression, but not necessarily in the same proportions. For a material having negative magnetostriction, the happenings resulting from tension or compression are reversed with respect to the happenings for a material having positive magnetostriction. The situation is further complicated by the fact that the happenings depend not only in degree but also in sign on the intensity of magnetization and stress and depend further on the order in which magnetization and stress are applied. In alloys, the magnitude and sign of magnetostrictive effects depend on the proportions of the materials. For example, in the commonly used nickel-iron alloys, the 50Ni-50Fe alloy has nearly the maximum strain sensitivity, whereas the 80Ni-20Fe

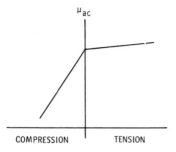

COMPRESSION | TENSION

Figure 8-1 Typical magnetostriction characteristic for nickel-iron alloys used in Draper Laboratory.

alloy has nearly zero strain sensitivity (change in induction/stress) for small stress. However, the elastic limit of the 80Ni alloy is easily exceeded, and for large stress it may give more trouble than the 50Ni alloy. For the nickel contents and flux densities commonly used in Draper Laboratory designs, Figure 8-1 gives a representative permeability variation.

The strains introduced by potting are due primarily to the manner in which the material sets during curing and thus are of a relatively permanent nature, but they can also cause magnetic materials to become very temperature sensitive owing to differential expansions and contractions that cause aggravation or relief of strains that already exist. Unless a potting compound is completely cured under properly controlled conditions, polymerization can continue for years, thus causing the effective impedances of magnetic suspension circuits and the flux patterns in the magnetic materials to change gradually with time. Flux densities are different in different parts of the magnetic structure and sometimes may contain biasing components; likewise, strain distribution may be very nonuniform and unsymmetrical, and thus a flux pattern may be considerably disturbed. The effect on permeability depends on whether ordinary alternating-current permeability or reversible (incremental) permeability is meant. The former can increase while the latter can decrease in the same circumstances.

While the primary effects of strain relate to changes in permeability and hysteresis of ferromagnetic materials, the resistivities of the materials are relatively less affected. Nevertheless, significant changes in eddy-current patterns can occur as an accompaniment of changes in flux-density patterns. For this reason, and more especially because of the usually undesirable effects of eddy currents in causing excessive heating and demagnetization at high frequencies, ferrite cores sometimes are used. Extremely thin laminations give very difficult handling problems during annealing, stacking, and bonding and give rather low stacking factors. Eddy-current effects in properly chosen ferrites can be essentially nil, though hysteresis may be substantial, and disaccommodation effects that are inconsequential in ferromagnetic materials may be troublesome. Ferrites have relatively low saturation flux densities, which at low frequencies would lead to relatively larger structures for given centering stiffness than are required if ferromagnetic material is used, but at high frequencies the ferrite structure can be the smaller. Manufacturing variations in ferrites generally are larger than for well-established ferromagnetic materials; and, being pressed mixtures of powdered oxides, ferrites are less likely to be as homogeneous as well-established ferromagnetic materials, so that specifications need to be imposed on suppliers for tolerances on permeability, disaccommodation, density, and porosity. Ferrites do not introduce heat-treatment problems, as do ferromagnetic materials, but the same precautions must be taken for avoiding the introduction of strains as discussed for ferromagnetic materials. The ferrites introduce a handling and mounting problem apart from the avoidance of strains, owing to brittleness. The only processing operations are light machining for mounting on a shaft or for finishing gap surfaces, which owing to hardness of the material generally is grinding, as light as possible.

For use in precision apparatus, the most important items in the handling and processing of magnetic materials are the correct annealing of ferromagnetic materials and the minimizing of the introduction of strains in either ferromagnetic or ferrite materials by the handling, machining, mounting, bonding, or potting process. The handling, machining, and mounting problems call largely for ingenuity and expertise of the designers and technicians, but the heat-treating, bonding, and potting in particular call for the advice of specialists in the materials and procedures best suited for these processes.

8-4 Eddy-Current Effects[51,52]

Eddy currents are detrimental owing to their contribution to core loss, to distortion of flux-density distribution, to demagnetizing action, and to delay of flux change when such delay is undesirable.

The magnetic suspensions developed at the Draper Laboratory are operated at frequencies in or somewhat above the audio range and currently at frequencies high enough so that for the nickel-iron laminations used the core loss is for practical purposes essentially all eddy-current loss. Eddy currents create a phenomenon known as magnetic skin effect, whereby the flux tends to crowd toward the surfaces of each lamination. This crowding increases with increases in lamination thickness, magnetic permeability, electrical conductivity, and source frequency. Rather than examine the entire classical electromagnetic approach to this problem, which is treated in detail in readily available literature, only those aspects that concern magnetic-suspension design are considered, namely, lamination thickness and magnetic saturation.

The flux density $\mathscr{B}_{ms}$ at either surface of a lamination of thickness $2s$, is related to the flux density $\mathscr{B}_{mc}$ at the center of the lamination by

$$\mathscr{B}_{ms} = \mathscr{B}_{mc} \cosh\left[2\pi \sqrt{\frac{\mu f}{\rho}}(1 + j1)s \right]. \tag{8-1}$$

These flux densities may be complex amplitudes, or rms values, on the assumptions that variation is sinusoidal in time, and that permeability μ and resistivity ρ are constant. Further, flux density $\mathscr{B}_{ms}$ is related to average peak flux density $\mathscr{B}_{ma}$ across the lamination by

$$\mathscr{B}_{ms} = 2\pi \sqrt{\frac{\mu f}{\rho}}(1 + j1)s\, \mathscr{B}_{ma} \coth\left[2\pi \sqrt{\frac{\mu f}{\rho}}(1 + j1)s \right]. \tag{8-2}$$

If

$$\frac{\mathscr{B}_{ms}}{\mathscr{B}_{ma}} = M_r + jM_x,$$

then

$$\mathscr{B}_{ms} = \mathscr{B}_{ma}\sqrt{M_r^2 + M_x^2}\,\Big/ \tan^{-1}\frac{M_x}{M_r}. \tag{8-3}$$

The parameters M_r and M_x are shown in Figure 8-2 as functions of the magnitude of the hyperbolic angle,

$$|\psi s| = 2\pi s\sqrt{\frac{2\mu f}{\rho}} = 4.44\tau\sqrt{\frac{\mu f}{\rho}},$$

in which

$$\tau = 2s,$$

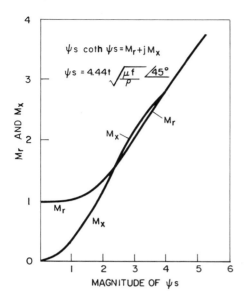

Figure 8-2 Chart of skin-effect parameters.

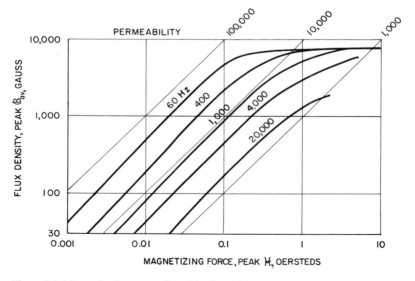

Figure 8-3 Magnetization curves for a 14-mil 80Ni-20Fe material.

the thickness of the lamination. As the curves indicate, $\mathscr{B}_{ms}$ differs in phase from both $\mathscr{B}_{mc}$ and $\mathscr{B}_{ma}$.

Theoretically, for a linear material, the value of $\mathscr{B}_{ms}$ corresponding to a $\mathscr{B}_{ma}$ based on the assumption that the magnetic material is linear can be computed from Equation (8-2); but, as readily can be appreciated, in reality $\mathscr{B}_{ms}$ may be much larger than $\mathscr{B}_{ma}$ so that the portions of the laminations near their surfaces may be saturated, and the desired total flux therefore is not achieved. For example, if at zero frequency a flux density of $\mathscr{B} = 2000$ gauss is desired in a 80Ni-20Fe material, a magnetizing force $\mathscr{H} = 0.04$ oersted is required, as may be estimated from Figure 8-3 by using the curve for 60 hertz. The normal permeability of this material is about 50,000 gauss/oersted, and its resistivity is about 50 microhm-centimeters. Hence, at 1000 hertz and for a 14-mil lamination, using cgs units,

$$\psi s = 4.44 \times 14 \times 2.54 \times 10^3 \sqrt{\frac{50 \times 10^3 \times 10^3}{50 \times 10^{-6} \times 10^9}} = 4.98,$$

so that Figure 8-2 gives

$$M_r = 3.5, \quad M_x = 3.5;$$

and if an average peak flux density $\mathscr{B}_{ma} = 2000$ gauss across the lamination is desired, to maintain the same total flux as at 60 hertz, the peak flux density at the surface of the lamination becomes

$$\mathscr{B}_{ms} = 2000\sqrt{(3.5)^2 + (3.5)^2} = 9900 \text{ gauss}$$

on the assumption that the material is linear, and it would require

$$\mathscr{H}_m = \frac{9900}{50} \times 10^3 = 0.198 \text{ oersted.}$$

But actually this indicated surface flux density is beyond the saturation limit of the material. According to Figure 8-3, to maintain an average peak flux density $\mathscr{B}_{ma} = 2000$ gauss at 1000 hertz requires $\mathscr{H}_m \approx 0.2$ oersted with an alternating-current permeability $\mu_{ac} \approx 10,000$, but the peak surface flux density $\mathscr{B}_{ms} \approx 6500$ gauss, obtained by following the $\mathscr{H} = 0.2$-oersted line to the 60-hertz curve. At the surface the eddy currents can have no demagnetizing effect, so that the average flux density at relatively low frequencies may be taken as essentially equal to the surface flux density at higher frequencies. In the example given, the surface flux density is over three times the average flux density, and the flux density at the center of the lamination must be substantially below the average flux density. This situation indicates rather inefficient use of the magnetic material. A practical estimate for a desirable lamination thickness can be made by use of Lord Rayleigh's formula for equivalent depth of penetration of the flux below the lamination surface:

$$d = \frac{1}{2}\sqrt{\frac{\rho}{\mu f}} = 2000\sqrt{\frac{\rho}{\mu_r f}} \text{ mils.} \tag{8-4}$$

The first form is correct for any consistent system of units, whereas the second form requires ρ to be in microhm-centimeters, μ_r to be relative permeability, and f to be in hertz. This depth is the depth the flux would occupy if it were distributed uniformly. Hence the lamination thickness should be

$$\tau \approx 2d,$$

which corresponds to $\psi s = \sqrt{2}$, $M_r = 1.08$, $M_x = 0.65$, and

$$\frac{\mathscr{B}_{ms}}{\mathscr{B}_{ma}} = \sqrt{(1.08)^2 + (0.65)^2} = 1.26,$$

indicative of only modest nonuniformity of distribution. Use of a thinner lamination can provide more flux with the same magnetomotive force and the same weight of material, the same flux with less magnetomotive force and the same weight of material, or the same flux with the same magnetomotive force and less weight of material. In addition to reduction of demagnetizing effects, reduction of lamination thickness of course decreases heating due to eddy currents.

The alternating-current magnetization curves for laminations are obtained under optimum conditions and hence do not indicate the deterioration that can result from bonding and final mounting and machining, due to inter-laminar conductance and the introduction of strains that influence eddy-current and hysteresis effects and permeability. Hence the designer is well advised to produce his own curves, such as illustrated in Figure 8-4, on final bonded, mounted, and machined cores.

In pulse-operated suspension systems, such as treated in Chapter 9, time delays are of special interest. The flux density and total flux buildup in a lamination subsequent to application of a step of magnetizing force are

$$\frac{\mathscr{B}_{xt}}{\mathscr{B}_{\infty}} = 1 - \frac{4}{\pi}\left(e^{-pt} \sin\frac{\pi x}{2s} + \frac{e^{-9pt}}{3} \sin\frac{3\pi x}{2s} + \frac{e^{-25pt}}{5} \sin\frac{5\pi x}{2s} + \ldots \right) \tag{8-5}$$

and

$$\frac{\phi_{xt}}{\phi_\infty} = \frac{x}{s} - \frac{8}{\pi^2}\left[\left(1 - \cos\frac{\pi x}{2s}\right)e^{-pt} + \left(1 - \cos\frac{3\pi x}{2s}\right)\frac{e^{-9pt}}{9}\right.$$
$$\left. + \left(1 - \cos\frac{5\pi x}{2s}\right)\frac{e^{-25pt}}{25} + \ldots\right], \tag{8-6}$$

in which $\mathscr{B}_{xt}$ and ϕ_{xt} are, respectively, the flux density at depth x and total flux within depth x at time t, $\mathscr{B}_\infty$ and ϕ_∞ are, respectively, the ultimate values of $\mathscr{B}_{xt}$ and ϕ_{xt}, and

$$p = \frac{\pi\rho}{16\mu s^2}.$$

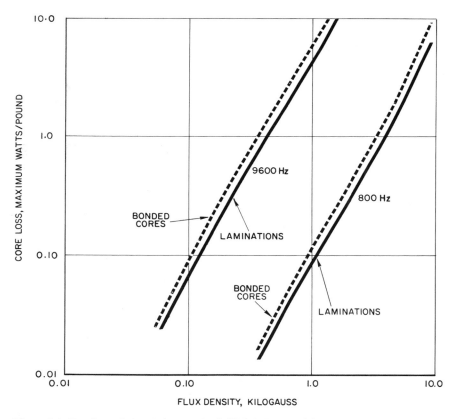

Figure 8-4 Core losses in bonded cores, 6-mil 50Ni-50Fe material.

Table 8-2 can serve as a guide in the selection of material and lamination thickness. The fractions were computed by use of Equation (8-6) with $x = s$ and, of course, can readily be extended to include other times and other materials.

Table 8-2. Fraction of Ultimate Flux Attained after 0.1 Millisecond

Materials	$\tau = 14$ mils	$\tau = 10$ mils	$\tau = 7$ mils	$\tau = 5$ mils
Allegheny 4750	0.21	0.30	0.38	0.59
Monimax	0.39	0.54	0.76	0.93
High-silicon steel	0.74	0.91	0.99	1.00

The high-permeability materials are inherently slow to respond unless their resistivities also are quite high. The response of high-resistivity ferrites is practically instantaneous compared with the responses of ferromagnetic materials.

8-5 Influences of Core Losses on Quality Factors

As for any ferromagnetic core device, the core losses and the demagnetizing effects of eddy currents reflect themselves into the apparent resistances and self-inductances of the operating coils, and hence into the apparent quality factors. For the magnetic suspension this situation becomes rather complicated because of the need for allocating the effects that are nonuniformly distributed over a multibranch magnetic circuit among the various coils on it. The effects for coils opposite the longer gaps are not necessarily the same as the effects for coils opposite the shorter gaps when the rotor is off center. Core losses reduce the effective inductance somewhat below the static, or direct-current, inductance and increase the effective resistance above the static, or direct-current, resistance, perhaps considerably, and thus reduce the effective quality factor. One way of approximating the situation is simply to use effective resistance and inductance measured with the rotor centered.

Experiments have shown that for nickel-iron alloys in certain operating regions (the reasons for which are explained presently) the effective Q_{Le} of the magnetic suspension is essentially constant for the respective coils, independent of rotor displacement, if flux densities do not go above the knee of the magnetization curve. For example, for a four-pole cylindrical suspension[53, 54]

$$Q_{Le} = \frac{\omega L_\ell I_1 + \omega N \Phi_1}{R_{1e} I_1}, \tag{8-7}$$

$$Q_{Le} = \frac{\omega L_\ell I_2 + \omega N \Phi_2}{R_{2e} I_2}, \tag{8-8}$$

$$Q_{Le} = \frac{\omega L_\ell I_3 + \omega N \Phi_3}{R_{3e} I_3}, \tag{8-9}$$

and

$$Q_{Le} = \frac{\omega L_\ell I_4 + \omega N \Phi_4}{R_{4e} I_4}; \tag{8-10}$$

these equations can be used to eliminate the RI terms in Equations (7-6) through (7-9). With

$$Q_{\ell e} = Q_{Le} - Q_{0e} \tag{8-11}$$

and

$$Q_{0e} = \frac{\omega N \Phi_0}{R_{0e} I_0} = \frac{\omega L_{0e}}{R_{0e}}, \tag{8-12}$$

in which the symbols with subscript zero indicate quantities corresponding to rotor centered, the solution can proceed as illustrated in Chapter 7 following Equation (7-9). Results for stiffness at zero displacement are shown in Figures 8-5a, 8-5b, 8-5c, and 8-6 for various proportions of leakage inductance. Details of the mathematics are given in the references.

A more general approach to the influence of core loss on the apparent quality factor Q_{Le} is made by viewing it as resulting from components that are assignable to the winding losses themselves and to the core losses, with the further approximations that for the nickel-iron alloys the eddy-current losses dominate, and that therefore with good approximation the hysteresis losses may be neglected, and that for ferrites the eddy-current losses may be neglected.

For thin nickel-iron alloy laminations, the eddy-current losses vary essentially as the square of the frequency for sinusoidally varying flux within the audio-frequency range, provided that the frequencies for which the losses are compared are not too widely separated. This frequency-squared relation is derivable on the assumption that flux-density distribution is uniform across the lamination and presumably should be true for any fixed flux-density distribution, but since flux-density distribution itself is a function of frequency, to keep the distribution essentially fixed would require some compensating changes in lamination thickness. Theoretically, as the frequency

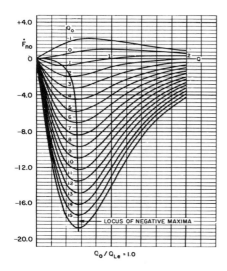

Figure 8-5a Normalized initial stiffness versus Q for various values of Q_0, Case (1) or (7) connection; $Q_0/Q_{Le} = 1.0$, Q_{Le} constant.

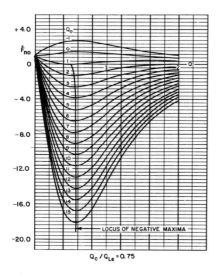

Figure 8-5b Normalized initial stiffness versus Q for various values of Q_0, Case (1) or (7) connection; $Q_0/Q_{Le} = 0.75$, Q_{Le} constant.

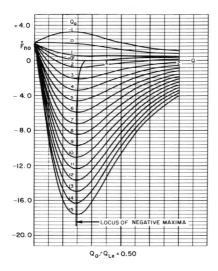

Figure 8-5c Normalized initial stiffness versus Q for various values of Q_0, Case (1) or (7) connection; $Q_0/Q_{Le} = 0.50$, Q_{Le} constant.

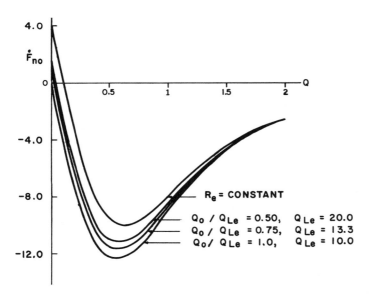

Figure 8-6 Normalized initial stiffness versus Q for $Q_0 = 10.0$ and various Q_0/Q_{Le} ratios, Case (1) or (7) connection.

becomes quite high, eddy-current losses should decrease and approach zero, owing to extreme magnetic skin effect, but for engineering purposes the frequency-squared relation has been shown to be quite tenable within the general limits stated. For very high resistivity ferrites, zero eddy-current loss and hysteresis loss in proportion to frequency is a very good approximation.

The quality factor of a coil may be generally defined as the ratio of reactive power input to active power input. In Figure 8-7(a), the true self-inductance of a coil is represented by L, the core losses can be imagined to occur in an equivalent shunt resistance r_i, and the winding resistance is represented by series resistance R_w. The parameters of the series equivalent circuit, Figure 8-7(b), are

$$R_i = \frac{r_i}{1 + \left(\dfrac{r_i}{\omega L}\right)^2} = \frac{r_i}{1 + Q_i^2} \approx \frac{r_i}{Q_i^2},$$

$$R_e = R_w + R_i,$$

and

$$L_e = \frac{L}{1 + \left(\dfrac{\omega L}{r_i}\right)^2} = \frac{L}{1 + \dfrac{1}{Q_i^2}} \approx L$$

for high Q_i, for example, $Q_i > 10$, depending on the approximation desired. Here

$$Q_i = \frac{r_i}{\omega L} = \frac{\omega L_e}{R_i} \approx \frac{\omega L}{R_i}$$

is the component of quality factor due to the core. The total quality factor of the coil is

$$Q_{Le} = \frac{\omega L}{R_w + R_i} = \frac{Q_w Q_i}{Q_w + Q_i} \tag{8-13}$$

because

$$\frac{1}{Q_{Le}} = \frac{R_w + R_i}{\omega L} = \frac{1}{Q_w} + \frac{1}{Q_i},$$

where Q_w is the component of quality factor due to the winding. In previous chapters, in which core loss is ignored, $Q_L = \omega L/R$ is the quality factor due

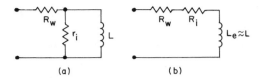

Figure 8-7 Equivalent circuits of coil: (a) with shunt core-loss resistance, (b) with series core-loss resistance.

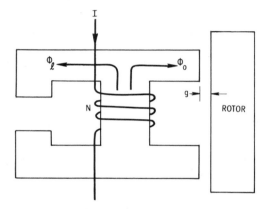

Figure 8-8 Equivalent magnetic circuit for separation of gap and leakage fluxes.

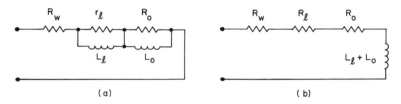

Figure 8-9 Equivalent electric circuits based on Figure 8-8.

to winding resistance, and R represents the winding resistance, all the loss being in the winding. Here R_w replaces R, and Q_w becomes Q_L when core loss is zero.

A further modification useful in application to the magnetic suspension can be made by means of the schematic magnetic circuit of Figure 8-8, representative of a pole pair of an eight-pole magnetic suspension, which allows for separation of leakage flux from the flux that crosses the gap between stator and rotor and produces the force between them. The equivalent electric circuits are shown in Figure 8-9 with equivalent resistances in which the eddy currents may be imagined to occur allocated separately to the leakage inductance and the component of inductance due to gap flux. On the basis of Figures 8-8 and 8-9,

$$R_i = R_\ell + R_0$$

and

$$Q_{Le} = \frac{\omega(L_\ell + L_0)}{R_w + R_\ell + R_0}.$$

Here R_ℓ is the series equivalent of r_ℓ, which may be imagined to account for core loss due to leakage flux, and R_0 is the series equivalent of r_0, which may be imagined to account for core loss due to gap flux with the rotor centered, on the assumption that though to a considerable extent the fluxes are superposed, the component losses may be viewed separately. Then

$$Q_w = \frac{\omega(L_\ell + L_0)}{R_w} = \frac{\omega L}{R_w} = Q_{w\ell} + Q_{w0},$$

$$Q_i = \frac{\omega(L_\ell + L_0)}{R_\ell + R_0} = \frac{\omega L}{R_i} = Q_{i\ell} + Q_{i0},$$

$$Q_{\ell e} = \frac{\omega L_\ell}{R_w + R_\ell + R_0} = \frac{\omega L_\ell}{R_e},$$

and

$$Q_{0e} = \frac{\omega L_0}{R_w + R_\ell + R_0} = \frac{\omega L_0}{R_e}.$$

Here Q_{w0}, the quality factor due to gap flux and winding resistance only, with the rotor centered, supersedes Q_0 in previous chapters. The last two equations give

$$Q_{Le} = Q_{\ell e} + Q_{0e}.$$

In the preceding equations,

$$L = N^2(\mathscr{P}_\ell + \mathscr{P}_0) = L_\ell + L_0,$$

in terms of leakage permeance $\mathscr{P}_\ell$ and gap permeance $\mathscr{P}_0$, for rotor centered, and the quality factors are for some base frequency f_b. Hence, for nickel-iron cores, at some other frequency f,

$$Q_{wf} = Q_w \frac{f}{f_b} \tag{8-14}$$

and

$$Q_{if} = Q_i \frac{f_b}{f}, \tag{8-15}$$

for practical purposes, over the audio-frequency range, and

$$Q_{Lef} = \frac{Q_{wf} Q_{if}}{Q_{wf} + Q_{if}} = \frac{Q_w Q_i}{Q_w \dfrac{f}{f_b} + Q_i \dfrac{f_b}{f}}. \tag{8-16}$$

For ferrite cores, Equation (8-14) applies, but since hysteresis loss and hence r_i varies directly with frequency (on the assumption that hysteresis loss varies also in proportion to the square of the flux density),

$$Q_{if} = Q_i, \tag{8-17}$$

a constant corresponding to base frequency, and

$$Q_{Lef} = \frac{Q_w Q_i}{Q_w + Q_i \dfrac{f_b}{f}}. \tag{8-18}$$

If f_b is selected to make

$$Q_w = Q_i = 2Q_{Le}, \tag{8-19}$$

the relations simplify as shown in Table 8-3.

Table 8-3. Quality-Factor Relations Based on f_b That Gives $Q_w = Q_i$

Quality Factor	Nickel-Iron	Ferrite
Q_{wf}	$2\,Q_{Le}f/f_b$	$2\,Q_{Le}f/f_b$
Q_{if}	$2\,Q_{Le}f_b/f$	$2\,Q_{Le}$
Q_{Lef}	$2\,Q_{Le}/(f/f_b + f_b/f)$	$2\,Q_{Le}/(1 + f_b/f)$
Q_{Lef} for $\dfrac{f}{f_b} \ll 1$	$2\,Q_{Lef}f/f_b = Q_w$	$2\,Q_{Le}f/f_b = Q_w$
Q_{Lef} for $\dfrac{f_b}{f} \ll 1$	$2\,Q_{Le}f_b/f = Q_i$	$2\,Q_{Le} = Q_i$
$Q_{Lef}(\text{max})$	Q_{Le}	$2\,Q_{Le}$

The condition of Equation (8-19) makes Q_{Lef} for nickel-iron cores a maximum at frequency f_b, but Q_{Lef} for high-resistivity ferrites does not peak but theoretically approaches its maximum asymptotically as f approaches infinity. The quality-factor relations as derived are not valid if interwinding capacitance is appreciable or if core losses are so large that fictitious resistances R_e and R_0 are so large that the use of self-inductance $L_l + L_0$ in Figure 8-9(b) is not a tolerable approximation.

If the rotor is displaced from its centered position, which means that in Figure 8-8

$$g = g_0(1 \pm x_n),$$

the resulting situation can be studied by resolving the quality factors into components that arise from the leakage and gap fluxes, respectively. For example,

$$Q_w = Q_{wl} + Q_{wg}$$

and

$$\frac{1}{Q_i} = \frac{1}{Q_{il}} + \frac{1}{Q_{ig}}.$$

With the rotor displaced,

$$Q_{wg} = \frac{Q_{w0}}{1 \pm x_n},$$

because self-inductance changes inversely with gap length, and

$$Q_{ig} = Q_{i0}(1 \pm x_n)$$

for the same reason. For $Q_{w0} = \omega L_0/R_w$ resistance R_w is constant, whereas for $Q_{i0} = r_i/\omega L_0$ resistance r_i is taken as constant. Here Q_{w0} and Q_{i0} are the quality factors when the rotor is centered. Hence, in general, and if Q_{w0} and Q_{i0} are taken at base frequency f_b,

$$Q_{wf} = \left(Q_{w\ell} + \frac{Q_{w0}}{1 \pm x_n}\right)\frac{f}{f_b} \tag{8-20}$$

and

$$\frac{1}{Q_{if}} = \left[\frac{1}{Q_{i\ell}} + \frac{1}{Q_{i0}(1 \pm x_n)}\right]\frac{f}{f_b} \tag{8-21}$$

for nickel-iron, and

$$\frac{1}{Q_i} = \left[\frac{1}{Q_{i\ell}} + \frac{1}{Q_{i0}(1 \pm x_n)}\right] \tag{8-22}$$

for high-resistivity ferrites, which give

$$Q_{Lef} \approx Q_{\ell f} + \frac{Q_{w0}Q_{i0}}{\dfrac{Q_{w0}}{(1 \pm x_n)}\dfrac{f}{f_b} + Q_{i0}(1 \pm x_n)\dfrac{f_b}{f}} \tag{8-23}$$

for nickel-iron, and

$$Q_{Lef} \approx Q'_{\ell f} + \frac{Q_{w0}Q_{i0}}{\dfrac{Q_{w0}}{1 \pm x_n} + Q_{i0}(1 \pm x_n)\dfrac{f_b}{f}} \tag{8-24}$$

for high-resistivity ferrites. Here $Q_{\ell f}$ and $Q'_{\ell f}$ are small equivalent leakage quality factors that are complicated functions of all the parameters involved. In general, $Q_{w\ell}$ is small and $Q_{i\ell}$ is large in comparison with either Q_{w0} or Q_{i0}, so that if $Q_{w\ell}$ and $1/Q_{i\ell}$ are neglected in Equations (8-20), (8-21), and (8-22), equivalent leakage quality factors $Q_{\ell f}$ and $Q'_{\ell f}$ do not appear. If, further, $Q_{w0} = Q_{i0} = 2Q_{0e}$ at base frequency f_b, then for gap flux only,

$$Q_{gef} \approx \frac{2Q_{0e}}{\dfrac{1}{(1 \pm x_n)}\dfrac{f}{f_b} + (1 \pm x_n)\dfrac{f_b}{f}} \tag{8-25}$$

for nickel-iron, and

$$Q_{gef} \approx \frac{2Q_{0e}}{\dfrac{1}{(1 \pm x_n)} + (1 \pm x_n)\dfrac{f_b}{f}} \tag{8-26}$$

for high-resistivity ferrites. At base frequency f_b, Equations (8-25) and (8-26) are practically constant over substantial ranges of x_n, which is in agreement with the assumptions made in the derivation based on Equations (8-7) through (8-10).

For either class of material at relatively low frequencies, $f/f_b \ll 1$,

$$Q_{gef} \approx \frac{2Q_{0e}}{1 \pm x_n}\frac{f}{f_b} \approx \frac{Q_0}{1 \pm x_n}; \tag{8-27}$$

for nickel-iron at relatively high frequencies, $f/f_b \gg 1$,

$$Q_{gef} \approx 2Q_{0e}(1 \pm x_n)\frac{f_b}{f}; \tag{8-28}$$

and for ferrite at relatively high frequencies,

$$Q_{gef} \approx 2Q_{0e}(1 \pm x_n). \tag{8-29}$$

The input impedance of the winding for a pole pair at rotor displacement x is

$$Z_{xf} = 2\left(R_{xf} + j\omega L_l + j\frac{\omega L_0}{1 \pm x_n}\right) = j2\omega L_l + 2R_{xf}\left[1 + j\frac{\omega L_0}{R_{xf}(1 \pm x_n)}\right],$$

in which $2R_{xf}$ is the effective input resistance of the windings on a pole pair. But for nickel-iron, if $Q_{w0} = Q_{i0}$ at frequency f_b,

$$Q_{gef} = \frac{\omega L_0}{R_{xf}(1 \pm x_n)} = \frac{\omega_b L_0}{\left[\dfrac{1}{(1 \pm x_n)}\dfrac{f}{f_b} + (1 \pm x_n)\dfrac{f_b}{f}\right]R_w},$$

because at base frequency f_b winding resistance R_w now is half the total effective resistance R_e, so that

$$R_{xf} = \left[\frac{f^2}{f_b^2}\frac{1}{(1 \pm x_n)^2} + 1\right]R_w; \tag{8-30}$$

and for high-resistance ferrites,

$$Q_{gef} = \frac{\omega L_0}{R_{xf}(1 \pm x_n)} = \frac{\omega_b L_0}{\left[\dfrac{1}{1 \pm x_n} + (1 \pm x_n)\dfrac{f_b}{f}\right]R_w},$$

so that

$$R_{xf} = \left[\frac{f}{f_b}\frac{1}{(1 \pm x_n)^2} + 1\right]R_w. \tag{8-31}$$

In deriving the relations that show the influence of gap length, the permeability of the magnetic materials is assumed to be infinite, or, practically, the reluctances of the paths through magnetic materials are supposed to be negligible compared with the gap reluctances. The range over which gap length can be changed, within the reasonable validity of the equations, therefore is limited. As the gap is made shorter the reluctances of magnetic materials become significant, and as it is made longer fringing becomes significant. The effects of frequency and displacement are shown in Figure 8-10, which really gives graphical illustration of Equations (8-25) and (8-26).

For nickel-iron cores,

$$\tan \beta = \frac{f_b}{f}(1 \pm x_n) = \frac{Q_{if}}{2Q_{0e}}(1 \pm x_n) = \frac{Q_{ig}}{2Q_{0e}}$$

and

$$\cot \beta = \frac{f}{f_b}\frac{1}{(1 \pm x_n)} = \frac{Q_{wf}}{2Q_{0e}}\frac{1}{(1 \pm x_n)} = \frac{Q_{wg}}{2Q_{0e}},$$

so that

$$Q_{gef} = \frac{2Q_{0e}}{\tan \beta + \cot \beta}, \tag{8-32}$$

which is in agreement with Equation (8-25) and is a maximum when

$$\tan \beta = \cot \beta = 1,$$

or $\beta_0 = 45°$. With the understanding that x_n may be either positive or negative, the relation that gives the optimum Q_{gef} for nickel-iron cores is

$$u = \frac{f}{f_b} = 1 + x_n, \tag{8-33}$$

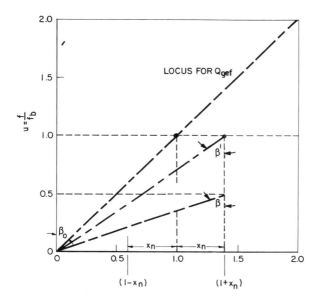

Figure 8-10 Frequency ratio u versus normalized gap length $1 \pm x_n$.

in which u is the normalized frequency f/f_b, or

$$x_n = u - 1, \tag{8-34}$$

obtained by differentiating Equation (8-25) with respect to u or with respect to x_n. The line for $\beta_0 = 45°$ is the locus of combinations of frequency and gap length that keep Q_{gef} at its maximum of Q_{0e}. The line for a general angle β is for other combinations of frequency and gap length, which result in smaller Q_{gef}, inversely proportional to the sum of the tangent and cotangent. In Equations (8-25) and (8-32), Q_{0e} is a constant based on frequency f_b and gap length g_0, and x_n is based on g_0.

For high-resistivity ferrite cores, Equation (8-32) may be written

$$Q_{gef} = \frac{2Q_{0e}}{\tan \beta + \cot \beta'}, \tag{8-35}$$

in which

$$\cot \beta' = \frac{1}{1 \pm x_n}.$$

This equation is in agreement with Equation (8-26). Differentiation of Equation (8-26) with respect to x_n gives

$$x_n = \sqrt{u} - 1 \tag{8-36}$$

for the displacement that gives maximum Q_{gef}. However, this relation gives

$$\text{maximum } Q_{gef} = Q_{0e}\sqrt{u} \tag{8-37}$$

for the maximum with respect to gap length. To understand this relation and recognize that it does not contradict the idea that theoretically Q_{gef} should approach Q_{if} as a limit as f increases, it should be reviewed as follows. If for some specific frequency ratio $u = u_1$, Equation (8-36) is inserted into Equation (8-26), Equation (8-37) results. If it now is written in the form

$$Q_{gef} = \frac{2Q_{0e}}{\dfrac{1}{\sqrt{u_1}} + \dfrac{\sqrt{u_1}}{u}},$$

obviously u now can be increased indefinitely by increase in frequency, so that the equation approaches

$$Q_{gef} = 2Q_{0e}\sqrt{u_1} = Q_{if}$$

corresponding to x_{n1}, which is the core-loss component of the quality factor for the gap length $g_0 + x_{n1}$ at very high frequency, in accordance with Equation (8-29). Further, at each new increase of frequency, matching values of Equation (8-36) can be inserted into Equation (8-26), so that for each increase in frequency, new values $x_{n2}, x_{n3}, x_{n4}, \ldots$ can be found to give new gap lengths that theoretically maximize Q_{gef} for each new frequency. But, practically, in this process the frequency eventually becomes so high that the relations are upset by capacitive effects, and the gap becomes so long that it cannot be treated as a "short" gap. For fixed gap length, Q_{gef} has no maximum with respect to frequency but simply approaches $2Q_{Le} = Q_i$ as frequency increases, aside from the influence of capacitance effects.

For high-resistivity ferrites, the β line in Figure 8-10 still is illustrative of the influence of winding loss, but the β' line illustrative of the influence of hysteresis loss must be separate and always terminates on the $u = 1$ line at the same x_n at which the β line terminates. When the two lines are coincident and their upper terminal moves along $u = 1$, the sum of tangent and cotangent becomes a minimum of 2 when $x_n = 0$; $Q_{wg} = Q_{ig}$, and Q_{ge} becomes a maximum for $u = 1$. If the upper terminal of the β line moves along a line of constant $u \neq 1$ as a locus, as the upper terminal of the β' line moves along $u = 1$, then the sum of tangent and cotangent becomes a minimum other than 2 at a value of $x_n \neq 0$; $Q_{wg} = Q_{ig}$, and Q_{ge} becomes a maximum for the new value of u, but all at different values than for $u = 1$, $x_n = 0$.

For an existing device, the frequency f_b at which $Q_{wg} = Q_{ig}$ for the given gap length g can be determined by measuring R_w and determining R_{ef} and $\omega(L_t + L_0)$ at a known frequency f by a simple resonance test. Then Q_{wf} and Q_{Lef} can be computed. For nickel-iron cores,

$$Q_{Le} = \frac{Q_{wf}}{2} \frac{f_b}{f} = \frac{Q_{Lef}}{2}\left(\frac{f}{f_b} + \frac{f_b}{f}\right),$$

(8-38)

obtained by use of Table 8-3. Hence

$$\frac{f}{f_b} = \sqrt{\frac{Q_{wf}}{Q_{Lef}} - 1}.$$

(8-39)

For high-resistivity ferrite cores,

$$Q_{Le} = \frac{Q_{wf}}{2} \frac{f_b}{f} = \frac{Q_{Lef}}{2}\left(1 + \frac{f_b}{f}\right),$$

(8-40)

also obtained by use of Table 8-3. Hence

$$\frac{f}{f_b} = \frac{Q_{wf}}{Q_{Lef}} - 1.$$

(8-41)

In either case, f_b can be found for the particular device. For a nickel-iron device, f_b is not only the frequency at which $Q_{wg} = Q_{ig}$, but it is the frequency at which Q_{gef} or Q_{Lef} is a maximum for gap length g, on the assumption that leakage inductance is essentially independent of gap length and frequency. For a ferrite device, f_b is merely the frequency at which $Q_{wg} = Q_{ig}$. Now if gap length g is changed and corresponding new values of Q_{wf} and Q_{Lef} are computed, new values of f_b can be computed (for either the nickel-iron or the ferrite device) for which new values of $Q_{wg} = Q_{ig}$ are determined. If for a high-resistivity ferrite the gap length is changed, then the new frequency at which the effective quality factor Q_{Lef} is unchanged may be obtained from

$$\frac{1}{1 + x_n} + (1 + x_n)\frac{f_b}{f} = 2,$$

which gives

$$u = \frac{f}{f_b} = \frac{(1 + x_n)^2}{2x_n + 1}. \tag{8-42}$$

At this frequency, $Q_{wgf} \neq Q_{igf}$. In Equation (8-42), x_n is limited to $x_n > -\frac{1}{2}$, and u is limited to $u > 1$.

In the relations involving change in gap length, if x_n is interpreted in the usual way for a single-axis suspension, then one pair of gaps increases in length and the other pair correspondingly decreases, and the quality factors apply to the circuits of the respective sides, for which the values of x_n are equal but of opposite sign. Likewise the equations would apply to the individual pole pairs of Figure 7-5 in which the pole pairs are not coupled. However, if x_n is interpreted as indicating an increase or decrease in all gap lengths, with the suspended member centered, then the equations apply equally to the quality factors of all circuits. In fact, the relations can apply to the design of reactors generally, entirely apart from the special case of magnetic suspensions.

Other graphical displays* of the effects of gap length and frequency on quality factor are in Figures 8-11, 8-12, and 8-13, in which the relations are plotted in somewhat more detail on log-log scales.

Figure 8-11 relates to nickel-iron cores. The diagonals and the lower curve are normalized from Equations (8-14), (8-15), and (8-16) for the condition $Q_w = Q_i$ for $u = 1$, with the aid of Table 8-3, and Q_{wf}/Q_{Lef} is plotted from the ratio of Equations (8-14) and (8-16) or could be obtained directly from Equation (8-38). This latter curve gives an indication of the influence of winding losses in establishing the quality factor. A scale of x_n can be added to Figure 8-11 by utilizing Equation (8-34). This scale means that if g_0 is changed to $g_0 + x$, the result on any of the curves is the same as if u had been changed from 1 to the value of u above the value of x_n, except that leakage inductance must be neglected. This matching of x_n and u on the two scales also means that if the gap length is changed by an amount corresponding to x_n, the frequency ratio must change from 1 to u, or if the frequency ratio is changed from 1 to u, the gap length must be changed by an amount corresponding to x_n, to maintain equality of Q_{wgf} and Q_{igf}. The same information can be obtained by recognizing that the ordinate scale of quality ratios may serve also as a "gap multiplier" scale, to be read by entering the $Q_{wf}/2Q_{Le}$ curve at some frequency ratio u. The gap multiplier is the number $m = 1 + x_n$ by which the

---

*These graphical displays, and the analytical work leading to them, are the development of G. A. Oberbeck and are believed by the authors to be original.

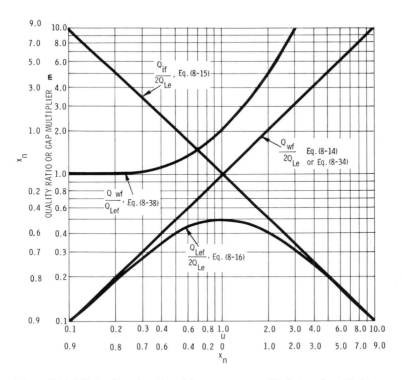

Figure 8-11 Effects of gap length and frequency on quality factors for nickel-iron cores.

gap length g_0 must be multiplied to obtain a new gap length $g_0 + x$ that keeps $Q_{wgf} = Q_{igf}$; for nickel-iron with eddy-current losses dominating m is the same as u. In fact, the x_n scale can just as well be placed beside the gap-multiplier scale as beside the u scale. Attention is called again to the fact that this equality of Q_{wgf} and Q_{igf} means also that they remain constant and maximum Q_{Le} remains constant, all at the values corresponding to base frequency f_b.

Figure 8-12 relates to high-resistivity ferrite cores. The 45° diagonal curve is obtained from Equation (8-14) as before. The $Q_{if}/2Q_{Le}$ curve is obtained from Equations (8-17) and (8-19) and is perfectly flat. The $Q_{Lef}/2Q_{Le}$ curve is obtained from Equation (8-18) for the condition $Q_w = Q_i$ for $u = 1$, with the aid of Table 8-3, and Q_{wf}/Q_{Lef} is obtained from the ratio of Equations (8-14) and (8-18) or could be obtained directly from Equation (8-40). This latter curve gives an indication of the influence of winding losses in establishing the

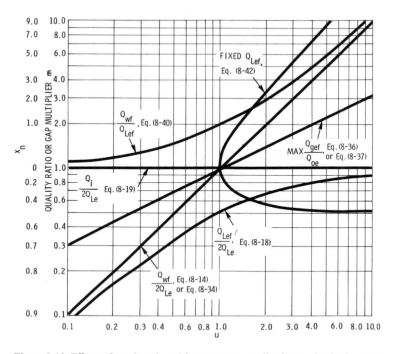

Figure 8-12 Effects of gap length and frequency on quality factors for ferrite cores.

quality factor of a coil on a high-resistivity ferrite core. The fixed Q_{Lef} curve is a plot of Equation (8-42). A scale of x_n can be added to Figure 8-12 also, in the same manner as for Figure 8-11, but preferably beside the gap-multiplier scale, and then it applies only to the additional curves of maximum Q_{gef}/Q_{0e} and fixed Q_{Lef}. The curve of Q_{gef}/Q_{0e} has a slope of $\frac{1}{2}$ on the log-log plot and can be interpreted also as a locus of $Q_{wgf} = Q_{ig}$. If this curve is entered at some frequency ratio u, the number m by which g_0 must be multiplied to give gap length $g_0 + x$ that makes $Q_{wgf} = Q_{ig}$ at the new frequency can be read from the gap-multiplier scale, and the corresponding normalized change in gap length can be read from the x_n scale. Conversely, if this curve is entered at some normalized gap-length change x_n, or the corresponding gap multiplier m, the frequency ratio that maintains $Q_{wgf} = Q_{ig}$ can be read from the u scale. Attention is called again to the fact that this equality of $Q_{wfg} = Q_{ig}$ does not mean that they remain constant at the values corresponding to base frequency f_b. If the curve of fixed Q_{Lef} is entered at some frequency ratio u, the number by which g_0 must be multiplied to keep Q_{Lef} fixed can be read from the gap-multiplier scale. Conversely, if this curve is entered at some normalized gap-length change x_n, or the corresponding gap multiplier m, the frequency ratio that maintains Q_{Lef} fixed can be read from the u scale. Since $x_n > -\frac{1}{2}, m > \frac{1}{2}$.

The relations for influence of gap length and frequency on the quality factors can be generalized to relate to cores for which both eddy-current and hysteresis losses are significant or for hybrid structures containing materials in some of which eddy-current losses dominate and other materials in which hysteresis losses dominate. If P_{hn} and P_{en} are the hysteresis and eddy-current losses each normalized with respect to total core loss P_c, then

$$\frac{P_h + P_e}{P_c} = P_{hn} + P_{en} = 1. \tag{8-43}$$

When base frequency f_b is adjusted so that $R_w = R_{i0}$ for gap length g_0, then if all core loss is eddy-current loss,

$$R_{igf} = \frac{u^2 R_w}{(1 \pm x_n)^2},$$

derived from Equation (8-30), and if all core loss is hysteresis loss,

$$R_{igf} = \frac{u R_w}{(1 \pm x_n)^2},$$

derived from Equation (8-31). Hence for a hybrid situation,

$$R_{igf} = \frac{u R_w}{(1 \pm x_n)^2}(P_{hn} + u P_{en}), \tag{8-44}$$

and if leakage flux is neglected,

$$Z_{xf} \approx 2R_w\left[1 + \frac{u(P_{hn} + u P_{en})}{(1 \pm x_n)^2} + j\frac{2u Q_{0e}}{1 \pm x_n}\right]. \tag{8-45}$$

If $x_n = 0$, leakage flux need not be neglected, and Equation (8-45) becomes

$$Z_f = 2R_w[1 + u(P_{hn} + u P_{en}) + j2u Q_{Le}]. \tag{8-46}$$

In this relation,

$$R_{if} = R_w u(P_{hn} + u P_{en}) \tag{8-47}$$

can be derived directly from core-loss relations if the assumption that hysteresis loss varies with the square of flux density is made, as before. Then, as always

$$\frac{Q_{wf}}{2Q_{Le}} = u,$$ (8-48)

but

$$\frac{Q_{if}}{2Q_{Le}} = \frac{u}{u(P_{hn} + uP_{en})} = \frac{1}{P_{hn} + uP_{en}}$$ (8-49)

and

$$\frac{Q_{Lef}}{2Q_{Le}} = \frac{u}{1 + u(P_{hn} + uP_{en})} = \frac{1}{\frac{1}{u} + P_{hn} + uP_{en}}.$$ (8-50)

If $P_{en} = 1$ and $P_{hn} = 0$, Equation (8-49) reduces to Equation (8-15) and Equation (8-50) reduces to Equation (8-16) for $Q_w = Q_i$ at $f = f_b$. If $P_{en} = 0$ and $P_{hn} = 1$, Equation (8-49) reduces to Equation (8-17) and Equation (8-50) reduces to Equation (8-18) for $Q_w = Q_i$ at $f = f_b$.

Figure 8-13 is a superposition of Figures 8-11 and 8-12. Curve 1 is the line of $Q_{wf}/2Q_{Le}$ as before, on both of the previous figures. The hatched region 2 is bounded by the limiting curves $Q_{if}/2Q_{Le}$ for eddy-current loss only and for hysteresis loss only, marked e and h, respectively. The speckled region 3 is bounded by the limiting curves $Q_{Lef}/2Q_{Le}$ for eddy-current loss only and for hysteresis loss only, marked e and h, respectively. The hatched region 4 is bounded by the limiting curves Q_{wf}/Q_{Lef} for eddy-current loss and for hysteresis loss only, marked e and h, respectively. Curve 5 is the line of maximum Q_{gef}/Q_{0e} taken from Figure 8-12. The dotted region 1-5 is a general area relating to gap multipliers, discussed presently. Intermediate lines in the specified areas can be plotted through use of Equations (8-49), (8-50), and the ratio of Equations (8-48) and (8-50). One such line, for $P_{hn} = \frac{2}{3}$ and $P_{en} = \frac{1}{3}$, is shown in each region, except region 1-5. Curve 6 is the curve of fixed Q_{Lef} taken from Figure 8-12.

If now x_n is reintroduced, Equation (8-45) gives

$$\frac{Q_{gef}}{2Q_{0e}} = \frac{1}{\frac{1 + x_n}{u} + \frac{P_{hn} + uP_{en}}{1 + x_n}},$$ (8-51)

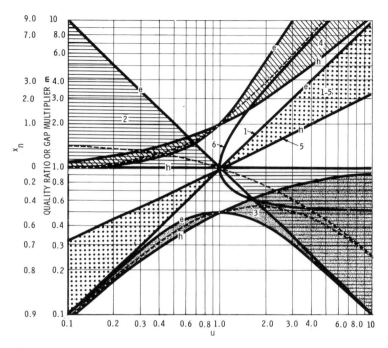

Figure 8-13 Superposition of Figures 8-11 and 8-12.

differentiation of which with respect to x_n gives

$$x_n = \sqrt{u(P_{hn} + uP_{en})} - 1 \tag{8-52}$$

for the value of x_n which makes Q_{gef} a maximum. It is the general condition for which $Q_{wgf} = Q_{igf}$. For $P_{hn} = 0$ and $P_{en} = 1$, this equation reduces to Equation (8-34), and for $P_{hn} = 1$ and $P_{en} = 0$, it reduces to Equation (8-36). The maximum value of Q_{gef} becomes

$$\text{maximum } Q_{gef} = \frac{Q_{0e}\sqrt{u}}{\sqrt{P_{hn} + uP_{en}}} \tag{8-53}$$

for constant u, and it reduces to Q_{0e} for $P_{hn} = 0$ and $P_{en} = 1$, and to $Q_{0e}\sqrt{u}$ for $P_{hn} = 1$ and $P_{en} = 0$, as previously shown. Equation (8-53) can be plotted on a chart such as Figure 8-13, but it cannot be interpreted against the x_n or gap-multiplier scales. Curve 1 for $Q_{wf}/2Q_{Le}$ serves also for entering the x_n or gap-multiplier scale of ordinates when $P_{hn} = 0$ and $P_{en} = 1$, and curve 5 for maximum Q_{gef}/Q_{0e} serves also for reading the x_n or gap-multiplier scale when $P_{hn} = 1$ and $P_e = 0$, as already explained. The dotted region 1-5 between these curves is an area in which corresponding points for x_n or the gap multiplier and the value of u lie for keeping the equality of Q_{wgf} and Q_{igf} when $P_{hn} \neq 0$ and $P_{en} \neq 0$, but no simple relation exists that makes the ordinate scale usable for general intermediate values of P_{hn} and P_{en}. However, the region can be used as a check area which covers the range of the gap multiplier, and any computed gap multiplier that falls outside of it must be in error, as is explained presently.

Differentiation of Equation (8-51) with respect to u gives

$$u = \frac{1 + x_n}{\sqrt{P_{en}}} \tag{8-54}$$

for the value of u that makes Q_{gef} a maximum. For $P_{en} = 1$ it reduces to Equation (8-34), and for $P_{en} = 0$ it becomes infinite, in agreement with previous findings. The maximum value of Q_{gef} becomes

$$\text{maximum } Q_{gef} = \frac{2Q_{0e}}{2\sqrt{P_{en}} + \dfrac{P_{hn}}{1 + x_n}} \tag{8-55}$$

for constant x_n, reduces to Q_{0e} for $P_{hn} = 0$ and $P_{en} = 1$, and approaches

$$2Q_{0e}(1 + x_n) = Q_i(1 + x_n)$$

as the frequency approaches infinity. If x_n were held constant at zero, Q_{gef} would approach Q_i as the frequency approaches infinity, as indicated on Figures 8-12 and 8-13.

If x_n is held constant at some finite value, the effect may be visualized as establishing a new base gap length

$$g_0' = g_0 + x_{n1},$$

for which if $P_{en} = 1$ and $P_{hn} = 0$, a new base frequency may be established,

$$f_b' = f_b(1 + x_{n1}),$$

for which the quality factors

$$Q_{w0}' = \frac{Q_{w0}}{1 + x_{n1}} \frac{f_b'}{f_b} = Q_{w0}$$

and

$$Q'_{io} = Q_{io}(1 + x_{n1})\frac{f_b}{f'_b} = Q_{io} = Q_{w0}$$

are kept equal and, in fact, unchanged. Then the curves may be used as plotted, but with

$$u' = \frac{f}{f'_b} = \frac{u}{1 + x_{n1}}.$$

If $P_{en} = 0$ and $P_{hn} = 1$, a new base frequency may likewise be established,

$$f''_b = f_b(1 + x_{n1})^2,$$

for which the quality factors

$$Q''_{w0} = \frac{Q_{w0}}{1 + x_{n1}}\frac{f''_b}{f_b} = Q_{w0}(1 + x_{n1})$$

and

$$Q''_{io} = Q_{io}(1 + x_{n1}) = Q_{w0}(1 + x_{n1})$$

are kept equal but change. The plotted curves may be used for this circumstance also, but with

$$u'' = \frac{f}{f''_b} = \frac{u}{(1 + x_n)^2}.$$

Now for a further change in gap length x'_{n1}, which must be based on g'_0, the procedures may be repeated, each time requiring new base frequencies. However, the procedure with x_1 always referred to the original g_0 and use of the original base frequency f_b seems simpler.

The intermediate curves in region 3 can be made relatively flat for a distance beyond $u = 1$ by appropriate selection of magnetic material or combination of materials. For example, if a second crossover is desired at

$$Q_{Lef}/2Q_{Le} = \tfrac{1}{2} \text{ for } u > 1, \text{ then}$$

$$\frac{1}{\dfrac{1}{u} + P_{hn} + uP_{en}} = \frac{1}{2},$$

or

$$\frac{1}{u} + P_{hn} + uP_{en} = 2. \tag{8-56}$$

For an assumed value of u, this equation can be solved simultaneously with Equation (8-43). For $P_{hn} = \frac{2}{3}$ and $P_{en} = \frac{1}{3}$, the crossover is at $u = 3$, as plotted. For $u = 5$, the crossover occurs for $P_{hn} = \frac{4}{5}$ and $P_{.en} = \frac{1}{5}$. Whereas such crossings theoretically can be found as u is increased indefinitely, the flatness deteriorates, P_{en} becomes so small as to be uncertain of achievement, and capacitive effects presently destroy the validity of the relations. According to Equations (8-54) and (8-55), with $x_n = 0$, the peaks are

$$\frac{Q_{Lef}}{Q_{Le}} = \frac{1}{P_{hn} + 2\sqrt{P_{en}}}$$

and occur at

$$u = \frac{1}{\sqrt{P_{en}}}.$$

Then

$$P_{hn} = \frac{u^2 - 1}{u^2}$$

and

$$P_{en} = \frac{1}{u^2}.$$

The utility of this scheme is not limited to magnetic-suspension applications but could be generally useful in reactor design, as could much of the work in this section.

Figures 8-14 through 8-21 give experimental evidence in verification of some of the theoretical work of this section.

Figure 8-14 shows curves of effective quality factor and effective resistance for a magnetic-suspension coil for which the magnetic core material is nickel-iron. The effective resistance at maximum quality factor is very close to double the winding resistance. Figure 8-15 gives the normalized quality factor of Figure 8-14, compared with the theoretical curve, on the log-log chart. If measurements of R_{ef} and Z_{ef} are made at a known frequency, and R_w is

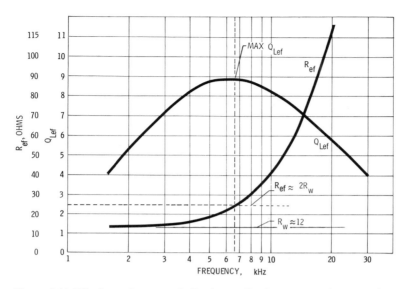

Figure 8-14 Effective resistance and effective quality factor versus frequency for a coil with nickel-iron core.

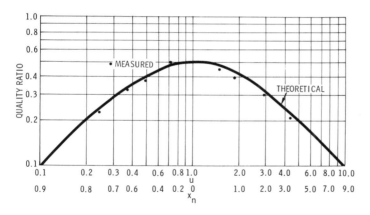

Figure 8-15 Normalized effective quality factor of Figure 8-14 versus normalized frequency compared with theoretical curve.

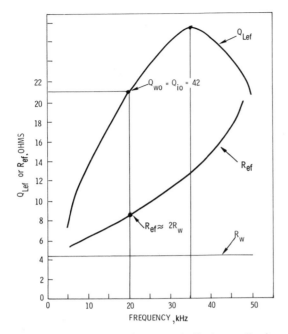

Figure 8-16 Effective resistance and effective quality factor versus frequency for a coil with a ferrite MN-31-M-A core.

measured, then Q_{Lef}, Q_{wf}, and $Q_{wf}/Q_{Lef} = R_{ef}/R_w$ can be computed. If this ratio is located on the Q_{wf}/Q_{Lef} curve of Figure 8-11 or 8-13 and the corresponding ratio u is noted, the base frequency f_b for the coil then can be computed. Then the theoretical $Q_{Lef}/2Q_{Le}$ curve can be interpreted with respect to the particular coil, and projection vertically to the Q_{wf}/Q_{Le} line, then horizontally to the ordinate scales gives the gap multiplier m and the corresponding x_n. If, as in this case, several sets of measurements are available, a scattering of base frequencies results, the average of which, if used with the data computed from measurements, gives a scattering of points with respect to the theoretical curve.

Figure 8-16 shows the effective quality factor and effective resistance for a magnetic-suspension coil having a ferrite MN-31-M-A core. When $Q_{w0} = Q_{i0}$, the effective resistance is very close to double the winding resistance. The quality factor Q_{Lef} does not increase indefinitely with increase of frequency owing to onset of capacitive effects.

Figure 8-17 shows the effective quality factor and effective resistance for a

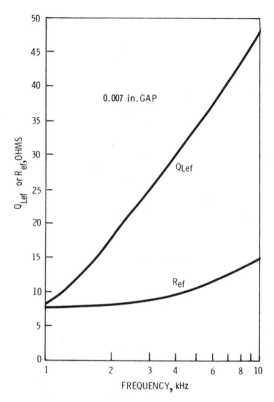

Figure 8-17 Effective resistance and effective quality factor versus frequency for a coil with a ferrite MN-60 core having 7-mil gaps.

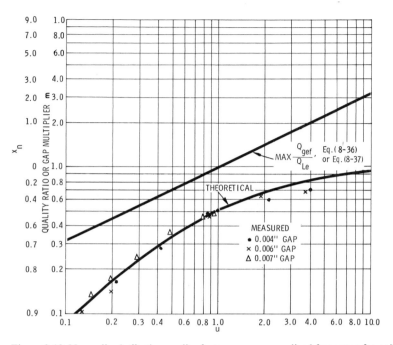

Figure 8-18 Normalized effective quality factor versus normalized frequency for coils with ferrite MN-60 cores having various gap lengths.

magnetic-suspension coil having a ferrite MN-60 core with 7-mil gaps. Figure 8-18 shows the normalized curve of $Q_{Lef}/2Q_{Le}$ for this coil and for coils of similar suspensions for which the gaps are 4 and 6 mils, compared with the theoretical curve. The points on this figure can be located from the experimental data by the same method as explained for Figure 8-15, using Figure 8-12 or 8-13. The gap multiplier m and the corresponding x_n can be found by projecting vertically to the maximum Q_{gef}/Q_{0e} line, then horizontally to the ordinate axis.

When both hysteresis and eddy-current losses are present in significant proportions, then measurements of R_{ef} must be made at two frequencies, so that P_{hn} and P_{en} can be determined. These measurements give

$$P_{hn} + u_1 P_{en} = \left(\frac{R_{ef1}}{R_w} - 1\right)\frac{1}{u_1}$$

and

$$P_{hn} + u_2 P_{en} = \left(\frac{R_{ef2}}{R_w} - 1\right)\frac{1}{u_2},$$

which if solved simultaneously with

$$P_{hn} + P_{en} = 1$$

and

$$\frac{u_1}{u_2} = \frac{f_1}{f_2}$$

give P_{hn}, P_{en}, and f_b. A simple way to determine Q_{wf}/Q_{Lef} is to measure power input P_f at frequency f, and power input P_0 at zero frequency, using equal currents. Then

$$\frac{Q_{wf}}{Q_{Lef}} = \frac{R_{ef}}{R_w} = \frac{P_f}{P_0}.$$

Then curves of $Q_{if}/2Q_{Le}$, $Q_{Lef}/2Q_{Le}$, and Q_{wf}/Q_{Lef} can be plotted versus u by computing points from Equations (8-49) and (8-50), and from the ratio of Equations (8-48) and (8-50). Any computed points that do not fall within the respective regions 2, 3, and 4 must be suspected of being in error. Whereas maximum Q_{gef}/Q_{0e} can be plotted from Equation (8-53), the gap multipliers m and the x_n values on the ordinate axis do not apply to it. However, unless this curve falls within the limits of region 1-5, it must be suspected of being in error.

Figure 8-19 gives a complete set of experimental curves for an eight-pole magnetic suspension having an Alfenol 16-percent aluminum-iron core, for which both hysteresis and eddy-current losses are significant. The various quality-factor ratios fall within the bounds of the regions defined on Figure 8-13. The winding resistance R_w and effective core resistance R_{if} are essentially equal at the base frequency $f_b = 700$ hertz where $Q_{wf} = Q_{if}$. Also, as a matter of interest, capacitance required for resonance of the circuit as a whole at frequency f is shown in terms of capacitance required for resonance at base frequency f_b.

The fitting of experimental data to the theoretical curves can be facilitated by using a transparency of Figure 8-13. Points computed from measured data can be plotted on two-decade log-log paper and placed under the transparency so that readings may be made from the various normalized theoretical curves,

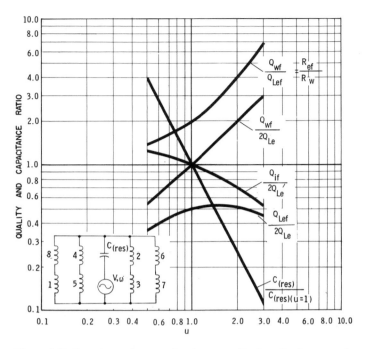

Figure 8-19 Experimental curves showing normalized quality factors and normalized resonant capacitance versus normalized frequency for eight-pole magnetic suspension with an Alfenol core.

including projections to determine the gap multiplier m and the corresponding x_n. Figure 8-20 is a set of curves of experimental data for quality factor Q_{Lef} versus frequency for eight-pole cylindrical suspensions having cores of Carpenter 49, ferrite, or 16-percent aluminum-iron plotted in rectangular coordinates, and Figure 8-21 is a normalizing of the same data on two-cycle log-log coordinates. Such plots can be obtained by fitting a log-log plot of the original data under a transparency of Figure 8-13.

Since the variations of the inherent properties of ferromagnetic materials and ferrites are difficult to control, as are changes in them caused by processing and machining, a design in which the resultant quality factor is dominated by Q_i is undesirable from the standpoint of manufacturing suspensions having reasonably predictable and uniform characteristics. In other words, design in the $u < 1$ region is desirable from this standpoint. However, from the standpoint of reliability of measurements to determine the parameters

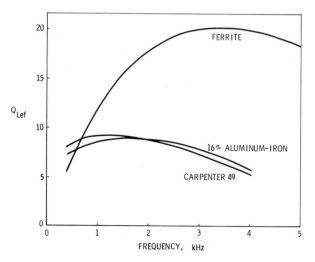

Figure 8-20 Experimental curves showing effective quality factors versus frequency for eight-pole cylindrical suspensions having cores of nickel-iron, ferrite, or aluminum-iron.

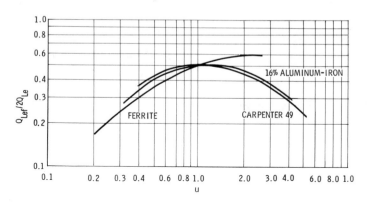

Figure 8-21 Normalized effective quality factors versus normalized frequency for data of Figure 8-20.

needed to compute curves such as are shown in Figures 8-11, 8-12, and 8-13, use of frequency in the $u > 1$ region, but not so high as to make capacitive effects significant, is desirable in order to separate the influences of winding and core. The small separation of the boundaries of region 3 for $u < 1$ illustrates the point.

8-6 Effects of Change in Quality Factor on Force Equation

As an example, force and stiffness equations of Case (1), series tuning with a voltage source, are generalized to take into account the effects of core loss, frequency base, and gap setting. In this generalization the function

$$U = u(P_{hn} + uP_{en})$$

is introduced. At the extreme of essentially all core loss arising from hysteresis, the function reduces to $U = u$; and at the extreme of practically all core loss arising from eddy currents, the function reduces to $U = u^2$. This generalization could have been introduced immediately in the previous section with some economy of equations and figures. However, building from the bounds of nickel-iron and high-resistivity ferrites, through Figures 8-11 and 8-12 to Figure 8-13, seems to add clarity.

First the generalized Case (1) equation is put in the form corresponding to Equation (2-7). In making this derivation, the original Equation (2-6) written for average force is used:

$$F_{Vs} = \frac{N^2 V^2 \mu A}{g_0^2} \left[\frac{1}{Z_1^2(1 - x_n)^2} - \frac{1}{Z_2^2(1 + x_n)^2} \right], \tag{2-6a}$$

but now

$$Z_1 = Z_{1xf} = 2R_w \left[\frac{U}{(1 - x_n)^2} + 1 \right] + j2R_w(U + 1)Q + j2 \frac{\omega L_0 x_n}{1 - x_n}$$

and

$$Z_2 = Z_{2xf} = 2R_w \left[\frac{U}{(1 + x_n)^2} + 1 \right] + j2R_w(U + 1)Q - j2 \frac{\omega L_0 x_n}{1 + x_n},$$

for $Q_{w0} = Q_{i0}$ at frequency f_b. Then

$$Z_{1xf}^2(1 - x_n)^2 = 4R_w^2 \left\{ \left(\frac{U}{1 - x_n} + 1 - x_n \right)^2 + [(U + 1)(1 - x_n)Q + Q_{wf} x_n]^2 \right\}$$

and

$$Z_{2xf}^2(1 + x_n)^2 = 4R_w^2\left\{\left(\frac{U}{1 + x_n} + 1 + x_n\right)^2 + [(U + 1)(1 + x_n)Q - Q_{wf}x_n]^2\right\}.$$

Substitution into Equation (2-6a) and introduction of the approximations $1 - x_n^2 \approx 1$ and $1/(1 \mp x_n) \approx 1 \pm x_n$ give

$$\frac{F_{Vs}}{F_V} \approx \frac{-4(U + 1)[QQ_{wf} - (Q^2 + 1) - U(Q^2 - 1)]x_n}{\{[U + 1 + (U - 1)x_n]^2 + [(U + 1)(1 - x_n)Q + Q_{wf}x_n]^2\}}, \quad (8\text{-}57)$$
$$\times \{[U + 1 - (U - 1)x_n]^2 + [(U + 1)(1 + x_n)Q - Q_{wf}x_n]^2\}$$

in which

$$F_V = \frac{N^2V^2\mu A}{4R_w^2g_0^2}$$

as for the idealized Case (1). Now if the equation is put in the form corresponding to Equation (5-1), for which the force base is the quiescent pull on one side of the block when it is centered,

$$\frac{F_{Vs}}{F_V} \approx \frac{\dfrac{-4[QQ_{wf} - (Q^2 + 1) - U(Q^2 - 1)]}{(U + 1)(Q^2 + 1)}x_n}{1 + \dfrac{\dfrac{-4[U - 1 + QQ_{wf} - (U + 1)Q^2]}{(U + 1)^2(Q^2 + 1)^2}x_n^2}{2(Q^2 + 1)\{U^2 + 1 + [(U + 1)Q - Q_{wf}]^2\}}}, \quad (8\text{-}58)$$
$$+ \dfrac{\{U^2 + 1 + [(U + 1)Q - Q_{wf}]^2\}^2}{(U + 1)^4(Q^2 + 1)^2}x_n^4$$

in which now

$$F_V = \frac{N^2V^2\mu A}{4R_w^2g_0^2(U + 1)^2(Q^2 + 1)}.$$

If Equation (8-58) is written for $Q = 1$, which is a common operating condition, it simplifies to

$$\frac{F_{Vs}}{F_V} \approx \frac{\dfrac{-2(Q_{wf} - 2)}{U + 1}}{1 + \dfrac{2[(U + 1)^2 - U(Q_{wf} + 1) + (Q_{wf} - 2)]}{(U + 1)^2}x_n^2}, \quad (8\text{-}59)$$
$$+ \dfrac{[U^2 + 1 + (U + 1 - Q_{wf})^2]^2}{4(U + 1)^4}x_n^4$$

for which the force base becomes

$$F_V = \frac{N^2 V^2 \mu^2 A^2}{8R_w^2 g_0^2 (U + 1)^2}.$$

As for Equation (5-1), the force base is a function of Q and, in contrast to the idealized Case (1), is also a function of U.

The normalized stiffness at zero displacement, derived from Equation (8-57), is

$$\dot{F}_{n0} \approx \frac{-4[QQ_{wf} - (Q^2 + 1) - U(Q^2 - 1)]}{(U + 1)^3 (Q^2 + 1)^2}, \tag{8-60}$$

on base

$$\dot{F}_V = \frac{N^2 V^2 \mu A}{4R_w^2 g_0^3};$$

whereas the normalized stiffness at zero displacement, derived from Equation (8-58), is

$$\dot{F}_{n0} \approx \frac{-4[QQ_{wf} - (Q^2 + 1) - U(Q^2 - 1)]}{(U + 1)(Q^2 + 1)}, \tag{8-61}$$

on base

$$\dot{F}_V = \frac{N^2 V^2 \mu A}{4R_w^2 g_0^3 (U + 1)^2 (Q^2 + 1)}.$$

For $Q = 1$, Equation (8-60) becomes

$$\dot{F}_{n0} \approx -\frac{(Q_{wf} - 2)}{(U + 1)^3} = -\frac{2(uQ_{0e} - 1)}{(U + 1)^3}, \tag{8-62}$$

and Equation (8-61) becomes

$$\dot{F}_{n0} \approx -\frac{2(Q_{wf} - 2)}{U + 1} = -\frac{4(uQ_{0e} - 1)}{U + 1}. \tag{8-63}$$

For the extremes of nickel-iron and high-resistivity ferrite cores, Equation (8-62) becomes

$$\dot{F}_{n0} \approx -\frac{2(uQ_{0e} - 1)}{(u^2 + 1)^3} \tag{8-64}$$

and

$$\dot{F}_{n0} \approx -\frac{2(uQ_{0e} - 1)}{(u + 1)^3},$$ (8-65)

respectively, and Equation (8-63) becomes

$$\dot{F}_{n0} \approx -\frac{4(uQ_{0e} - 1)}{u^2 + 1}$$ (8-66)

and

$$\dot{F}_{n0} \approx -\frac{4(uQ_{0e} - 1)}{u + 1},$$ (8-67)

for nickel-iron and high-resistivity ferrite cores, respectively. These expressions are evaluated for several frequency ratios u in Table 8-4.

Table 8-4. Stiffness at $x = 0$ as Influenced by Frequency Ratio u

Equation	$u \ll 1$	$u = 0.5$	$u = 1.0$	$u = 2.0$
(8-64)	$-Q_{wf} + 2$	$-0.51Q_{0e} + 1.02$	$-0.25Q_{0e} + 0.25$	$-0.032Q_{0e} + 0.016$
(8-65)	$-Q_{wf} + 2$	$-0.30Q_{0e} + 0.59$	$-0.25Q_{0e} + 0.25$	$-0.148Q_{0e} + 0.074$
(8-66)	$-2Q_{wf} + 4$	$-1.6Q_{0e} + 3.2$	$-2.0Q_{0e} + 2.0$	$-1.6Q_{0e} + 0.80$
(8-67)	$-2Q_{wf} + 4$	$-1.33Q_{0e} + 0.67$	$-2.0Q_{0e} + 2.0$	$-2.67Q_{0e} + 1.33$

To evaluate the actual stiffness, one must remember that f_b for the ferrite core that makes $Q_w = Q_i$ probably is much higher than the corresponding base frequency for a nickel-iron core and that Q_{wf} depends directly on actual frequency f, whereas Q_{0e} is fixed by the base frequency f_b for the particular device. Also, unless the devices under comparison are geometrically identical and have identical windings, the corresponding stiffness bases $\dot{F}_V$ are not the same.

The force and stiffness relations for the other cases, that use current sources, parallel circuit connections, or bridge circuits, can be developed by methods similar to the method used for Case (1) but are omitted and left for development by the reader who may have need of them, to avoid further large algebraic display.

In Figure 8-22, plots of Equation (8-57) are exhibited for comparison with Figure 2-4. For Figure 8-22, the substitution $2uQ_{Le} = Q_{wf}$ has been made, and the curves have been plotted for $2Q_{Le} = 10$ and $Q = 1$, for values of U

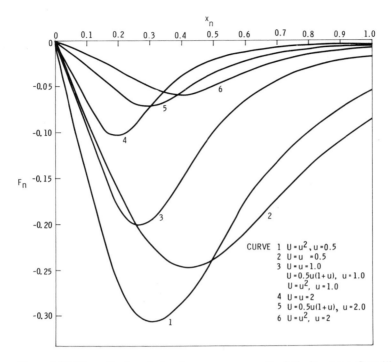

Figure 8-22 Normalized centering force versus normalized displacement, Case (1) connection, $Q_{Le} = 10$, $Q = 1$, for various proportions of hysteresis and eddy-current losses.

that correspond to u values for hysteresis loss only, equal hysteresis and eddy-current losses at base frequency, and eddy-current loss only. The curve for zero core loss, $U = 0$, and for $u = 1$ would be the same as the curve for $Q = 1$ in Figure 2-4, which should serve as the base for comparison.

In Figure 8-23, plots of Equation (8-60) are exhibited for comparison with Figure 2-13. For Figure 8-23 the conditions are the same as for Figure 8-22, except that, of course, $x = 0$ and the stiffness is plotted against Q. The curve for $U = 0$ and $u = 1$, would be the same as the curve for $Q_0 = 10$ in Figure 2-13, which should serve as the basis of comparison.

Figures 8-22 and 8-23 show that the core losses considerably reduce both the restoring-force level and the initial stiffness in comparison with those quantities for the ideal Case (1) suspension.

Figure 8-24 is a plot of experimental data of force versus displacement for a magnetic suspension having a high-resistivity ferrite core and therefore

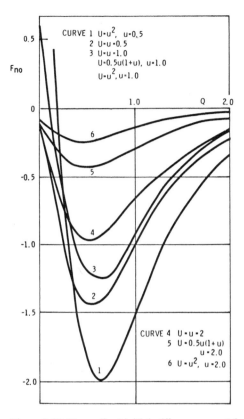

Figure 8-23 Normalized initial stiffness versus Q for curves of Figure 8-22.

negligible eddy-current loss. At each frequency, the capacitors were adjusted for half-power-point operation, and the voltage was adjusted to maintain fixed quiescent current, as shown in the associated tabulation. Theoretically, if the force relation is correctly given by Equation (8-59), this procedure should keep the force base essentially fixed, because then

$$Z_{1xf}^2 = Z_{2xf}^2 = 8R_w^2(u + 1)^2 = Z_{0f}^2$$

and

$$V_f^2 = I_0^2 Z_{0f}^2,$$

the curves of force should follow Equation (8-59), and the indicated force level

	V_{IN}	I_0	C_W	Q_{Lef}
10 kHz	0.80 V	70 mA	.0438 μF	10.3
20 kHz	1.06 V	70 mA	.0105 μF	17.5
40 kHz	1.90 V	70 mA	.0028 μF	22.9
70 kHz	3.11 V	70 mA	.0006 μF	19.2

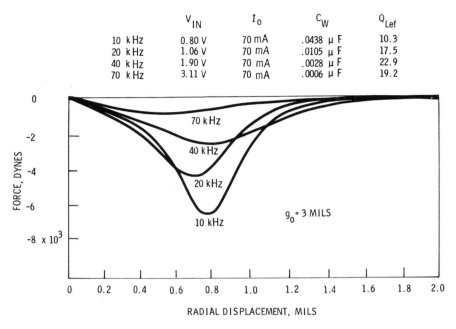

Figure 8-24 Measured centering force versus displacement for various frequencies for magnetic suspension with ferrite core, with same quiescent current for each curve, Case (1) connection; $Q = 1$.

should decrease progressively as frequency is increased. However, as study of the tabulation of Q_{Lef} indicates, stray capacitive effects must have started to contribute significantly to the impedances between 10,000 and 20,000 hertz and thereafter considerably influenced the force curves. The tuning action of the stray capacitance in effect shunted around the coils first increases the share of the total quiescent current received by the coils as frequency is increased but, ultimately, for further increase of frequency, decreases the share of the total quiescent current received by the coils. For example,

$$\frac{I_{coil}}{I_{total}} = \frac{\dfrac{1}{j2\omega C_{stray}}}{R_{ef} + j\left(\omega L - \dfrac{1}{2\omega C_{stray}}\right)} = \frac{1}{1 - 2\omega^2 L C_{stray} + j2\omega R_{ef} C_{stray}},$$

or

$$\left|\frac{I_{\text{coil}}}{I_{\text{total}}}\right|^2 = \frac{1}{(1 - 2\omega^2 L C_{\text{stray}})^2 + (2\omega R_{ef} C_{\text{stray}})^2}. \tag{8-68}$$

Here C_{stray} is the stray capacitance effectively in parallel with a coil pair, so that unless it has significant influence at relatively low frequencies,

$$\left(\frac{1}{2\omega C_{\text{stray}}}\right)^2 \gg R_{ef}^2 + (\omega L)^2,$$

or

$$[(2\omega R_{ef} C_{\text{stray}})^2 + (2\omega^2 L C_{\text{stray}})^2] \ll 1,$$

which means that then

$$(2\omega R_{ef} C_{\text{stray}})^2 \ll 1,$$

$$(2\omega^2 L C_{\text{stray}})^2 \ll 1,$$

and

$$\left|\frac{I_{\text{coil}}}{I_{\text{total}}}\right| \approx 1.$$

At base frequency, around 15,000 hertz, $R_{ef} = 2R_w$ and $\omega = \omega_b$, so that for half-power-point operation,

$$2R_w < \omega_b L,$$

and at any other frequency

$$R_w(u + 1) < u\omega_b L,$$

or in Equation (8-68), always

$$R_{ef} < \omega L.$$

Hence as ω increases the denominator of Equation (8-68) can decrease substantially below 1 but ultimately must become substantially greater than 1, so that if the total quiescent current always is set at 70 milliamperes, for the illustration of Figure 8-24, at a relatively low frequency it is essentially the conduction current in the coils, but as frequency is increased and the displacement current in the stray capacitance becomes significant, the quiescent conduction current in the coils theoretically increases to a maximum at relative frequency

$$u_p = \frac{3}{4} \frac{R_w^2}{(R_w^2 + \omega_b^2 L^2)} \pm \sqrt{\frac{1}{2(R_w^2 + \omega_b^2 L^2)} \left[\frac{L}{C_{\text{stray}}} - \frac{2\omega_b^2 L^2 R_w^2 - 5R_w^2}{2(R_w^2 + \omega_b^2 L^2)} \right]},$$

then it decreases to the initial quiescent (low frequency) I_0 again at relative frequency

$$u_0 = -\frac{R_w^2}{R_w^2 + \omega_b^2 L^2} \pm \sqrt{\frac{1}{R_w^2 + \omega_b^2 L^2} \left(\frac{L}{C_{\text{stray}}} - \frac{\omega_b^2 L^2 R_w^2}{R_w^2 + \omega_b^2 L^2} \right)}$$

and thereafter decreases further. The half-power-point setting here is for the entire circuit including stray capacitance and hence has not its usual significance in determining the performance of the suspension. In fact, even with the coil in parallel resonance with its shunt stray capacitance at relative frequency

$$u_R = -\frac{R_w^2}{R_w^2 + \omega_b^2 L^2} \pm \sqrt{\frac{1}{R_w^2 + \omega_b^2 L^2} \left[\frac{1}{2C_{\text{stray}}} - \frac{\omega_b^2 L^2 R_w^2}{R_w^2 + \omega_b^2 L^2} \right]}$$

a half-power-point setting gives a leading total current, but the coil currents nevertheless can give a stable suspension action. Here

$$u_p < u_R < u_0.$$

Though the constant current held is not coil current, the procedure serves to bring out the effect of stray capacitance. If the supply voltage had been held constant, the force curves would have been successively lower for increase in frequency, as for the ideal suspension, and the effect of stray capacitance would have been largely masked. If the values of Q_{Lef} tabulated on Figure 8-24 are normalized on Figure 8-12, the effects of stray capacitance are evident in the 20 to 40 kilohertz region and are very evident thereafter. For a ferromagnetic core, the effects of eddy currents most likely would take hold substantially before stray capacitive effects become significant as frequency is increased and thus largely mask the stray capacitance effects.

For any frequency setting, as the suspended member is displaced, the currents in coils adjacent to the opening gaps increase, reach a maximum, then decrease, and the currents in the coils adjacent to the closing gaps decrease, resulting in the force-displacement curves of Figure 8-24. All of this action can be represented reasonably well by complicated algebraic expressions written on the assumption that the effect of stray capacitance can be represented by lumped capacitances C_{stray} that shunt the coils, which of course is just a first approximation. Nevertheless, the description of the action given here, based on lumped shunt capacitances, though not mathematically exact, gives a qualitative explanation by means of which the principal effects of stray capacitance on the performance of a magnetic suspension can be visualized. The result is essentially a combination of series and parallel tuning.

8-7 Effect of Core Loss on Currents and Flux Densities

The manner of current change and accompanying flux density change with displacement when core loss is significant, compared with the manner of these changes when core loss is insignificant, is of interest. These changes are illustrated also for Case (1), the voltage source with series tuning.

When $x = 0$,

$$Z_{1f}^2 = Z_{2f}^2 = 4R_w^2(U + 1)^2(Q^2 + 1);$$

and when displacement is sufficient to produce resonance,

$$2R_w(U + 1)Q + \frac{2\omega L_0 x_n}{1 - x_n} = 0$$

for Z_{1xf}, or

$$2R_w(U + 1)Q - \frac{2\omega L_0 x_n}{1 + x_n} = 0$$

for Z_{2xf}. so that resonance occurs at

$$x_{nR} = \mp \frac{(U + 1)Q}{Q_{wf} - (U + 1)Q}.$$

Use of x_{nR} in Z_{1xf} or Z_{2xf} gives

$$Z_R^2 = \frac{4R_w^2\{U[Q_{wf} - (U + 1)Q]^2 + Q_{wf}^2\}^2}{Q_{wf}^4}$$

$$= \frac{R_w^2\{U[2uQ_{Le} - (U + 1)Q]^2 + 4u^2Q_{Le}^2\}^2}{4u^4Q_{Le}^4}.$$

Hence

$$\frac{I_0^2}{I_R^2} = \frac{\{U[2uQ_{Le} - (U + 1)Q]^2 + 4u^2Q_{Le}^2\}^2}{16u^4Q_{Le}^4(U + 1)^2(Q^2 + 1)},$$

which for $U = 0$ and $u = 1$, reduces to Equation 2-13. For nickel-iron cores,

$$\frac{I_0}{I_R} = \frac{u^2[2uQ_{Le} - (u^2 + 1)Q]^2 + 4u^2Q_{Le}^2}{4u^2Q_{Le}^2(u^2 + 1)\sqrt{Q^2 + 1}}$$

$$= \frac{1}{\sqrt{Q^2 + 1}}\left[1 - \frac{uQ}{Q_{Le}} + \frac{u^2 + 1}{4}\left(\frac{Q}{Q_{Le}}\right)^2\right];$$

and for high-resistivity ferrite cores,

$$\frac{I_0}{I_R} = \frac{u[2uQ_{Le} - (u + 1)Q]^2 + 4u^2Q_{Le}^2}{4u^2Q_{Le}^2(u + 1)\sqrt{Q^2 + 1}}$$

$$= \frac{1}{\sqrt{Q^2 + 1}}\left[1 - \frac{Q}{Q_{Le}} + \frac{u + 1}{4u}\left(\frac{Q}{Q_{Le}}\right)^2\right].$$

Values of I_R/I_0 for $2Q_{Le} = 10$ and $Q = 1$ are given in Table 8-5.

Table 8-5. Ratio of Resonant Current to Quiescent Current

Frequency Ratio →	$u = 0.5$	$u = 1.0$	$u = 2.0$
I_R/I_0 for nickel iron	1.54	1.72	2.18
I_R/I_0 for ferrite	1.70	1.72	1.73

The considerable increase in current with displacement when core losses are significant means that if the operation of the device is limited by approach to magnetic saturation as the suspended member is displaced, the quiescent current must be decreased as u is increased, and hence the stiffness at zero displacement and the maximum restoring force are correspondingly decreased.

At the longer gap, which has the higher flux density when the block of Figure 2-1 is displaced, the ratio of flux density at resonance to flux density at zero displacement is

$$\frac{\mathcal{B}_R}{\mathcal{B}_0} = \frac{I_R}{I_0}\frac{1}{1 + x_{nR}} = \frac{(U + 1)\sqrt{Q^2 + 1}}{\dfrac{U[2uQ_{Le} - (U + 1)Q]}{2uQ_{Le}} + \dfrac{2uQ_{Le}}{2uQ_{Le} - (U + 1)Q}}.$$

For nickel-iron cores,

$$\frac{\mathcal{B}_R}{\mathcal{B}_0} = \frac{(u^2 + 1)\sqrt{Q^2 + 1}}{\dfrac{u^2[2uQ_{Le} - (u^2 + 1)Q]}{2uQ_{Le}} + \dfrac{2uQ_{Le}}{2uQ_{Le} - (u^2 + 1)Q}};$$

and for high-resistivity ferrite cores,

$$\frac{\mathcal{B}_R}{\mathcal{B}_0} = \frac{(u + 1)\sqrt{Q^2 + 1}}{\dfrac{u[2uQ_{Le} - (u + 1)Q]}{2uQ_{Le}} + \dfrac{2uQ_{Le}}{2uQ_{Le} - (u + 1)Q}}.$$

Values of $\mathcal{B}_R/\mathcal{B}_0$ for $2Q_{Le} = 10$ and $Q = 1$ as well as the corresponding values of x_{nR} are given in Table 8-6.

Table 8-6. Ratio of Flux Density at Resonance to Quiescent Flux Density

Frequency Ratio →	$u = 0.5$	$u = 1.0$	$u = 2.0$
$\mathcal{B}_R/\mathcal{B}_0$ for nickel iron	1.16	1.40	1.63
x_{nR} for nickel-iron	0.33	0.25	0.33
$\mathcal{B}_R/\mathcal{B}_0$ for ferrite	1.19	1.40	1.47
x_{nR} for ferrite	0.43	0.25	0.176

The flux-density ratios in Table 8-6 of course are correct only before the onset of saturation. Hence, as mentioned for the current ratios, the higher the ratio, the smaller the quiescent flux density $\mathcal{B}_0$ must be for any allowable maximum flux density, and the weaker the suspension. As u becomes small or very large, x_{nR} reaches 1 at

$$u \approx \frac{Q_{Le}}{2Q} \pm \sqrt{\frac{Q_{Le}^2}{4Q^2} - 1}$$

for nickel-iron cores, and as u becomes small, x_{nR} reaches 1 at

$$u = \frac{Q}{Q_{Le} - Q}$$

for high-resistivity ferrite cores. With $2Q_{Le} = 10$ and $Q = 1$, x_{nR} becomes 1 at $u = 0.21$ or 4.79 for nickel-iron cores and at $u = 0.25$ for high-resistivity ferrite cores. Since $x_{nR} = 1$ means a closed gap, any value of u beyond these limits gives fictitious results, though mathematically x_{nR} increases to infinity and then becomes negative. The upper limit of u for nickel-iron cores probably is meaningless also, owing to the onset of capacitive effects.

Different assumptions for Q_{Le}, Q_{wf}, and Q of course give different results from the tabulations for both current ratios and flux-density ratios. With the current ratios available, the flux-density ratios can be found simply by dividing by $1 + x_{nR}$; or having the flux-density ratios, the current ratios can be found by multiplying by $1 + x_{nR}$, instead of using the long formulas in both instances.

The use of flux density at resonance is not in itself especially meaningful but gives a simple means of showing the trend of flux density with displacement. Flux density at resonance is not exactly maximum flux density at the gaps and pole faces, and depending on the design of the structure, the pole-face flux density may not be the bottleneck, as mentioned in the next section. Depending on the design of the structure, leakage fluxes may have substantial limiting influence.

8-8 Effects of Magnetic Saturation and Other Nonlinearities of the Magnetization Curve

The derivations in this monograph are based on the assumption that the reluctances of the gaps are so dominant in the magnetic circuit that the reluctances of the magnetic materials may be neglected. To make this assumption valid practically means that the parts of the magnetic circuit must be proportioned so that all of it is operated below the knee of the magnetization curves for the various materials over the range of displacement of the suspended body. Otherwise current distortions with high peaks, excessive core losses and winding losses with adverse heating effects, and premature force limitations are encountered. If miniaturization is an objective, one always has the problem of too much heat versus too much mass as the force demanded is increased.

For example, with the simple single-axis suspension, the gap flux density at the end having the longer gap must increase from the quiescent value to produce a restoring force. If with no displacement the quiescent flux density in the gaps is such that the flux density in the adjacent magnetic material already is beyond the knee of the magnetization curve, the core and copper losses already may be high relative to the heat-dissipating ability of the structure and would increase with displacement of the suspended member. Furthermore, with increased current at the end with the longer gap, the increase in flux could be very little, and likewise the decrease of flux at the other end could be very little, thus resulting in very little restoring force and a rather soft suspension, not governed by the relations derived herein.

For the more complex structure of a multipolar multiaxis suspension illustrated in Figure 8-25, the situation is somewhat similar but more complicated. Here the relative cross-sectional areas of the rotor at the pole faces, between slots, and below slots and the cross-sectional area of the stator normal to

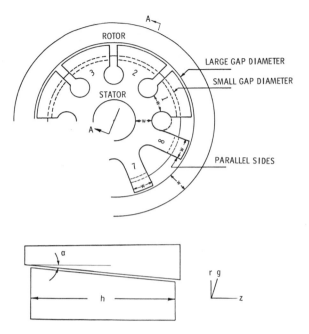

Figure 8-25 Illustration of tailoring of cross-sectional areas of magnetic circuit of rotor to avoid flux bottlenecks.

circumferential flux opposite the slot openings must be considered, with due regard to leakage fluxes, as well as gap fluxes, and different rotor and stator materials.

Examination of Figure 8-25 shows that in a path around poles 1 and 2, maximum flux density in either rotor or stator occurs at the side having the larger gap diameter, owing to the taper; that for the stator the maximum is either between or below the slots; and that for the rotor the maximum is circumferential, opposite the stator slot openings. When the rotor is displaced, this situation is emphasized in the parts nearer the longer gap. If any of these areas is a flux bottleneck, it limits the gap flux and hence limits the restoring force the suspension can develop. Under such circumstances, the force actually can be increased by reduction of polar areas as illustrated in Figure 8-25 for poles 7 and 8, which move provides more coil space also. Here all cross-sectional areas are essentially equal to wh, except as adjusted to allow for leakage fluxes, on the assumption that stator and rotor materials are the same. If the stator is ferrite and the rotor is nickel-iron, the rotor cross-sectional area can be much smaller than the stator cross-sectional areas.

When gap lengths are quite short compared with the lengths of the flux paths in the magnetic material or if the excitation is high, the nonlinear characteristics of the magnetic material influence the effective currents and flux densities and hence influence the force-displacement characteristics. This influence may result from magnetic saturation, as mentioned at the beginning of this section, or may result from the multivalued character of the hysteresis loop at any excitation level for relatively short gaps. The effects are different for different materials and for different magnetic circuit geometries and, therefore, are not readily subjected to general representation by test results or by analytical attack. Whereas some advantages may be gained by capitalizing on the nonlinear characteristics of materials in the design of magnetic suspensions, the subject requires more study, but as already noted, reliance on such characteristics does not give suspensions having reasonably predictable and uniform performance.

8-9 Magnetic Suspension Torques

Ideally the suspension forces should be purely radial or axial. As has been explained in Section 7-6, when a three-axis suspension is made by combination of two conical suspensions, radial and axial displacement may have coupling actions such that a radial displacement may generate a net axial force; and likewise axial displacement may cause changes in radial stiffnesses of the suspensions. Hence, even if geometry is perfect, torque can be generated about an axis normal to the axis of a suspension pair. If the center lines of the stator, rotor, and shaft axis are misaligned due to machining inaccuracies, spurious torques can be generated whether or not the suspension is a part of a three-axis pair. Small torques can arise also from irregularities in the machining of rotor and stator gap surfaces,[1] from inhomogeneous and non-isotropic properties of magnetic materials, from hysteresis and eddy-current effects, from magnetic disaccommodation, from motion of the float through the flotation fluid,[55] or from a combination of such causes. In general, these conditions are difficult to analyze, and separation of the various causes by measurement is difficult if not practically impossible, but some qualitative comments and analyses of simplified cases may be helpful.

The misalignment of a two-axis suspension is illustrated in Figure 8-26, in which rotor, stator, and shaft axes are not coincident and do not necessarily intersect. The misalignment of rotor and stator axes gives a net torque T_{ry} due

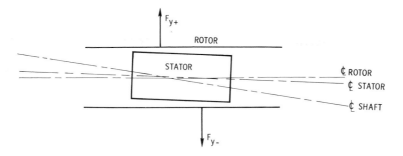

Figure 8-26 Illustration of misalignment of a two-axis suspension.

to the net force components F_y acting on the rotor parallel to the y axis. This couple acts in a plane determined by the y axis and the center line of the rotor. If the shaft axis does not lie in or parallel to that plane, torque T_{ry} has a component about the shaft or output axis

$$T_{0y} = T_{ry} \sin \phi_y,$$

in which ϕ_y is the angle that the output axis makes with the plane of T_{ry}. Likewise a torque T_{rx} may exist due to the net force components F_x acting on the rotor parallel to the x axis. This couple would act in a plane determined by the x axis and the center line of the rotor. If the shaft axis does not lie in or parallel to that plane, torque T_{rx} has a component about the shaft axis

$$T_{0x} = T_{rx} \sin \phi_x,$$

in which ϕ_x is the angle which the output axis makes with the plane of T_{rx}. The total torque about the shaft due to these misalignments then is

$$T_{0A} = T_{rx} \sin \phi_x + T_{ry} \sin \phi_y. \tag{8-69}$$

If free to turn, the rotor would tend to seek a position that makes Equation (8-69) zero.

The same relations apply in a somewhat more complicated way to each of two conical suspensions mounted on the same shaft to form a three-axis suspension, and if free to rotate, the assembly would seek a position for which the net torque of the two suspensions is zero. If in addition the "axial" pulls on the two suspensions are not aligned with the shaft axis, their axial components produce a tumbling torque, and their circumferential components with respect to the shaft axis produce torque about that axis.

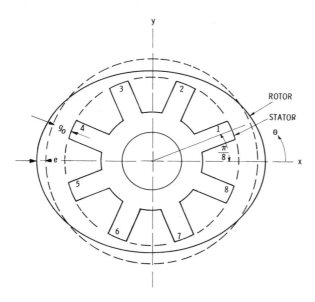

Figure 8-27 Illustration of elliptical rotor outside stator having circular outline.

Irregularities of machining of gap surfaces may result in reluctance variations that cause torque. A simple case is a slightly elliptical rotor, shown in Figure 8-27, outside a circular stator, with a TIR (Total Indicated Runout) of $2e$. For very small eccentricity e and for very small angular displacement θ of rotor with respect to stator, the gap lengths at the respective poles are

$$g_1 = g_5 = g_0 + e \cos 2\left(\frac{\pi}{8} - \theta\right),$$

$$g_2 = g_6 = g_0 + e \cos 2\left(\frac{3\pi}{8} - \theta\right),$$

$$g_3 = g_7 = g_0 + e \cos 2\left(\frac{5\pi}{8} - \theta\right),$$

and

$$g_4 = g_8 = g_0 + e \cos 2\left(\frac{7\pi}{8} - \theta\right).$$

The self-inductances of the paired coils neglecting coupling among the respective coil pairs (which is a good approximation when θ is small) are

$$L_{81} = 2L_\ell + \frac{(2N)^2 \mu A}{g_1 + g_8} = L_{45} = 2L_\ell + \frac{(2N)^2 \mu A}{2g_0 + \sqrt{2}e \cos 2\theta}$$

$$= 2L_\ell + \frac{2L_0}{1 + 0.707e_n \cos 2\theta}$$

and

$$L_{23} = 2L_\ell + \frac{(2N)^2 \mu A}{g_2 + g_3} = L_{78} = 2L_\ell + \frac{(2N)^2 \mu A}{2g_0 - \sqrt{2}e \cos 2\theta}$$

$$= 2L_\ell + \frac{2L_0}{1 - 0.707e_n \cos 2\theta},$$

in which

$$e_n = \frac{e}{g_0}.$$

The torque hence is

$$T = 2\left(\frac{I_{81}^2}{2} \frac{dL_{81}}{d\theta} + \frac{I_{23}^2}{2} \frac{dL_{23}}{d\theta}\right),$$

in which

$$I_{81}^2 = \frac{V^2}{Z_{81}^2},$$

$$I_{23}^2 = \frac{V^2}{Z_{23}^2},$$

$$Z_{81}^2 = 4R_w^2\left[1 + \left(Q - \frac{0.707Q_0e_n \cos 2\theta}{1 + 0.707e_n \cos 2\theta}\right)^2\right],$$

$$Z_{23}^2 = 4R_w^2\left[1 + \left(Q + \frac{0.707Q_0e_n \cos 2\theta}{1 - 0.707e_n \cos 2\theta}\right)^2\right],$$

$$\frac{dL_{81}}{d\theta} = \frac{2\sqrt{2}L_0e_n \sin 2\theta}{(1 + 0.707e_n \cos 2\theta)^2},$$

$$\frac{dL_{23}}{d\theta} = -\frac{2\sqrt{2}L_0e_n \sin 2\theta}{(1 - 0.707e_n \cos 2\theta)^2},$$

insertion of which in the torque equation gives

$$T \approx \frac{V^2 e_n^2 L_0 [Q(Q_0 - Q) - 1]}{R_w^2 (Q^2 + 1)^2} \sin 4\theta \tag{8-70}$$

if terms in the denominator containing e_n are neglected. This torque is a maximum when $\theta = \dfrac{\pi}{8}, \dfrac{3\pi}{8}, \dfrac{5\pi}{8}, \ldots,$.

The torques that arise from axial offset or misalignment or from gap-surface irregularities are in the class of reluctance torques caused by variation in gap dimensions. In the derivation of these torques, the magnetic materials are regarded as ideal. Inhomogeneous and nonisotropic magnetic materials can cause distortions in flux-density distributions that give the rotor preferred positions owing to small reluctance torques that arise from internal irregularities. These inhomogeneous and nonisotropic properties can cause unsymmetrical distribution of eddy currents, as can geometric dissymmetries and defects in interlamination resistance, which can give the rotor preferred positions owing to small induction torques. Hysteresis effects may cause residual torques if the unit is not demagnetized after erection. Demagnetization is facilitated by the heavy damping of the flotation fluid. During erection, the rotor is displaced from its unsupported position radially, axially, and rotationally, and the flux density in the gaps may be well above normal level. Hence residual fluxes can remain that cause rotor and stator to act as permanent magnets and cause small torques through attraction and repulsion. This action is troublesome especially in active, or servo-controlled, magnetic suspensions.

These small residual torques that cause the rotor to have preferred positions may be approximated in the form $T_M \sin n\theta$. In the preferred positions, the torque is zero, but it changes for rotation in either direction, giving an elastic restraint action. In intermediate positions, approximately halfway between preferred or zero-torque positions, the spurious torque is a maximum but changes very little with rotation. In some applications the change in spurious torque may be much more damaging than the presence of the spurious torque itself, which usually can be compensated in one way or another. Hence the quiescent operating point of the rotor may be set at maximum rather than zero spurious torque.

If the currents in the various coils of a magnetic suspension are not in time phase, a rotating field results that tends to make the rotor turn as an induction motor or as a hysteresis motor, or both. Such difference in time phase can occur temporarily if the rotor is displaced from its neutral position or can occur with the rotor in neutral position owing to capacitance trimming or other trimming of the electric circuits and to inherent slight inequities of resistance and inductances. In fact this principle is used in the design of auxiliary compensating windings for neutralizing the small spurious torques that arise from various sources.* These windings are placed on the stator, similar to a two-phase motor winding, usually to generate small hysteresis torque in a ferrite rotor.

In the high-precision applications of magnetic suspensions magnetic disaccommodation[56, 57] is troublesome. This phenomenon occurs in magnetic materials, especially ferrites, when the flux density is suddenly reduced and manifests itself as a slow change in alternating-current permeability. The change is very small and requires a long time for completion, perhaps minutes or even hours. Accommodation is the reverse of disaccommodation, caused by an increase of flux density. The time constants associated with accommodation are many times smaller than the time constants associated with disaccommodation. The magnitude of the change and the time constants associated with it are functions of the impurities of the materials, the magnitude of the step of flux-density change, and the level of magnetizing force at which the step is made. If such step occurs due to rotor displacement, either by rotation or translation, a spurious reluctance torque or force can be developed. The phenomenon is particularly troublesome with pulsed or square-wave operation.

All the spurious torques and torques discussed in this section are extremely small, but when accuracies in terms of parts per million are under consideration, they can be very significant. Whereas in general only the resultant error due to a variety of causes can be measured, some approximate analysis of the individual contributing causes is desirable to indicate the extreme care that must be taken in maintaining very close machining tolerances and in the selection and handling of materials.

8-10 Analogous Problems for Electric Suspensions

Most of the materials problems considered for magnetic suspensions have their parallels for electric suspensions: skin effects, eddy-current effects, dielectric losses, with their influences on quality factors, and so on. Dielectric

* A scheme developed by Howard L. Watson, an assistant director of the Draper Laboratory.

absorption phenomena are somewhat analogous to magnetic accommodation and disaccommodation. For the electric suspension, the magnetic materials problems are relegated to the external tuning inductors, where the troubles created are relatively minor.

For the electric suspension the material of principal concern is the dielectric, which also is the damping material and therefore is a liquid except for a thin skim of solid material deposited on one or both capacitor surfaces to provide insulation as protection against contact.

From the electrical standpoint, desirable properties of the dielectric are high breakdown strength, high dielectric constant, low dielectric losses, and low dielectric absorption. From the mechanical standpoint, desirable properties are constant viscosity, independent of temperature change, and density such as to make the float and contents appear substantially weightless to the magnetic suspensions, so that they can act substantially as trimming devices to keep the float off its bearings and nearly centered in the face of large accelerating forces and shocks. Further, the fluid should not undergo chemical changes owing to normal temperature cycling, dielectric stresses, or radiation encountered in space vehicles, nor should it interact with the capacitor-plate materials of the electric suspension or with the magnetic materials of the magnetic suspension. Unfortunately, the viscosity, dielectric constant, dielectric losses, and dielectric strength are quite sensitive to temperature change, so that close temperature control is necessary for high-precision devices.

Table 8-7. Properties of Dielectric Damping Fluids

	BTFE	CTFE	Units
Density	2.39	1.95	g/cm^3 at 125°F
Viscosity	4500.	2400.	cP at 125°F
Viscosity-temperature coefficient	5.8	5.0	%/°F at 125°F
Coefficient of cubic expansion	4.4×10^{-4}	4.4×10^{-4}	/°F
Specific heat		0.32	cal/g
Bulk modulus		0.12×10^{-6}	psi
Dielectric constant	3.40		at 1 kHz, 77°F
	3.26	2.82	at 1 kHz, 130°F
Dissipation factor		0.0009	at 130°F
Dielectric strength	500.	500.	V/mil for 0.1 in. at 130°F
Newtonian behavior	yes	yes	above 115°F

Typical fluids used in the Charles Stark Draper Laboratory over a period of years are fluorocarbons, polychlorotrifluorethylene (CTFE) and polybromotrifluorethylene (BTFE), the former being a polymer of the basic monomer $(C_2ClF_3)_x$ and the latter being a polymer of the monomer $(C_2BrF_3)_x$. They can be processed so that the batch viscosities can be varied from 50 to 3000 centipoises at a particular temperature. These fluids are just two of many possibilities. Their properties are given in Table 8-7. These data are from various sources and should be regarded merely as nominal values.

8-11 Summary
At relatively low frequencies, with relatively low flux densities in the magnetic material, and with gaps sufficiently long to have reluctances of paths through magnetic material negligible in comparison with gap reluctances, performance of magnetic suspensions can be very close to the behavior derived on idealized bases if machining tolerances are closely held and the working of the materials is kept to a minimum. Similar conditions apply to electric suspensions with respect to frequency, applied voltage, gap lengths, and dimensional tolerances. Operating temperature is of importance, especially for ferrites, dielectrics, and damping fluids, and in general for highly precise performance, control of ambient temperature is essential. As frequency increases, eddy-current effects become troublesome in ferromagnetic materials, ultimately even for the thinnest practical laminations, with resulting heat loss, deterioration in quality factor, distortion of flux-density distribution, and demagnetization. Hysteresis losses increase for all magnetic materials, with attendant deterioration in quality factor. Dielectric losses increase, with attendant deterioration in quality factor, dielectric constant, and dielectric strength. Shunting effects of stray capacitances become troublesome. Actually, the use of higher and higher frequency is not the designer's idea, but the 100-fold increase during the past 20 years is a situation largely forced by the advantages high-frequency power provides for other components in the vehicles in which the suspensions are used. Actually, for a force device, high frequency is basically disadvantageous rather than advantageous. Likewise, from the standpoint of precision and predictability of performance, high flux density is undesirable, or in general any design that depends significantly on properties of materials is undesirable from that standpoint. However, relatively high flux densities may have to be tolerated to keep the device within specified bounds of size and weight. As usual, the situation presented to the designer has no one best answer. The methods of normalization, as embodied in Figures 8-11, 8-12, and 8-13, can be of considerable aid to the designer in making his choices.

9 ACTIVE AND HYBRID SUSPENSIONS

In recent years, the development of solid-state electronic components and integrated circuits has revived interest in the active type of suspension with which the Draper Laboratory development began, owing to the great reduction in bulk of the control apparatus thus made possible in contrast with the use of vacuum tubes and conventional circuit components. Whereas twenty years ago the apparatus for the active control of the suspension of a gyro float might have required space approximating the size of a shoe box, the equivalent now can be placed within a cubic inch of space, or less. In contrast with the passive type of suspension, the active type of suspension has the possible advantages of substantially increased stiffness, increased rapidity of response, and decreased energy demand, with essentially no difference in space required for auxiliary circuitry.

9-1 Historical Background[1-3,7,8,32,33,46]

As indicated in Chapter 1, the magnetic suspension as originally developed in the Draper Laboratory was an active, or servo-controlled, type. It used a four-pole stator and a cylindrical rotor, as shown in Figure 7-1 for the passive type of suspension. The circuitry for the original scheme,[2] as developed about 1952, is shown in Figure 9-1 schematically, divorced from details of the electronic circuitry. With the rotor centered, the suspension coils carried equal direct currents and equal alternating currents. The directions of these currents were such that no direct or alternating voltage appeared between opposite corners of the coil mesh. The alternating currents served merely to provide position signals to indicate displacement of the rotor along the x axis or y axis, and the direct currents served to provide the electromagnetic forces. Displacement along the y axis, for example, would increase the inductances of coils 1 and 2 and decrease the inductances of coils 3 and 4, so that these changes, reacting with the associated capacitances, caused an alternating voltage to appear between top and bottom corners of the coil mesh. This voltage signal was taken to a preamplifier through blocking capacitors, then demodulated and applied to a direct-current amplifier with polarity such that the current supplied by it would be unbalanced in favor of coils 3 and 4, thus tending to reduce the y-axis displacement. The capacitors associated with the alternating-current source served also to block direct current from it. Similar circuitry controlled the restoring force along the x axis, and the combination

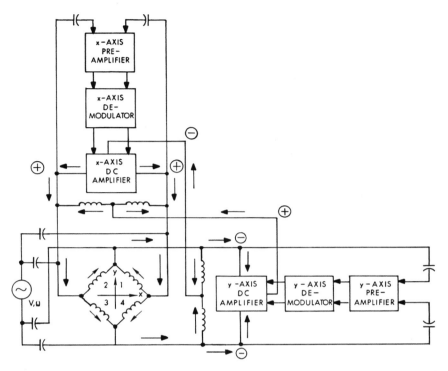

Figure 9-1 Essentials of circuitry for original Gilinson-Scoppettuolo mesh-connected magnetic suspension.

of the two force components tended to restore the rotor from displacement in any radial direction. This circuit had the advantage of requiring that only four leads be brought out but had the disadvantage of requiring one conventional "B-plus" power supply and one unconventional "B-minus" power supply for the direct-current feedback amplifiers and the disadvantage of having no common point for grounding.

Several months later the circuit was modified to the "star" connection[3] of coils of Figure 9-2 in which the common point of the star could be the common or ground terminal for both the alternating-current excitation circuit and the direct-current push-pull amplifier, for which the x'- and y'-axis control circuitry is shown. Similar circuitry exercised control with respect to the z axis. This modification accomplished some economy of apparatus, including the discarding of the awkward "B-minus" power supply, and also made

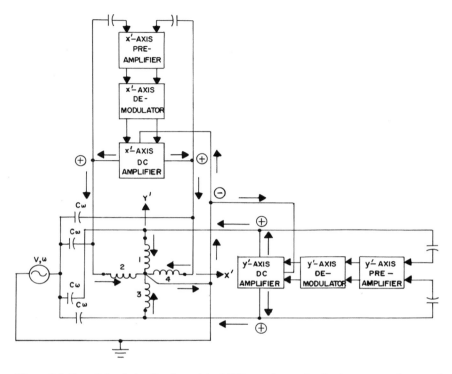

Figure 9-2 Essentials of circuitry for original Gilinson-Scoppettuolo star-connected magnetic suspension.

the force and control axes coincident with the Microsyn polar axes. Here also the alternating voltage was intended merely to provide position signals to the preamplifier, but by unbalance of the bridge circuit formed by coils 1 and 3 and associated capacitors, as described in detail in Section 2-7 for the single-axis suspension. The output of the preamplifier was demodulated and fed to the direct-current amplifier, and depending on the input to this amplifier, its output favored coil 1 or coil 3.

During the development of this active circuitry, the effect of the alternating currents used in the star connection for producing the displacement signals on the stiffness of the suspension was under study. Settings of the capacitances in the bridge circuit that placed the inductance-resistance-capacitance paths beyond the resonant point at zero displacement were observed to make the suspension stiffer than capacitance settings that placed the paths below resonance. In fact with the beyond-resonance settings and the direct-current supply disconnected, the rotor remained suspended nevertheless, and further study showed this action to be the basis of the passive type of suspension as analyzed in the earlier parts of this monograph as Case (1), variations of which until recently have dominated the Draper Laboratory development. Later study showed that the mesh connection also performed passively as described briefly in Section 7-5. Thus these original suspensions really were hybrid suspensions, involving both active and passive types of control. In some later developments this superposition has been deliberately incorporated to bolster the stiffness of the passive alternating-current suspension acting alone.

9-2 Quiescent Current; Time Sharing and Multiplexing; Pulsing

The passive type of magnetic suspension always has current in its coils when the suspended body is in its centered or zero-net-force position. When the suspended body is displaced, the coil currents increase or decrease automatically from the quiescent value in accordance with impedance changes in the tuned circuitry, so as to produce restoring forces. The energy dissipated in the windings and in the cores corresponds essentially to the quiescent currents, the additional energy dissipation caused by the current excursions being relatively small. If the quiescent currents could be eliminated, the energy requirements could be considerably reduced, corresponding only to the transient currents accompanying displacement of the suspended body from its quiescent position. For an active suspension, the quiescent position corresponds to the signal center.

The active type of magnetic suspension may have substantial quiescent current, relatively small quiescent current, or none. In this sense suspension operation is analogous to Class A, B, or C operation of electronic amplifiers[58] and in fact is guided by the manner of operation of the electronic components that form the driving source. For Class A push-pull operation, the restoring force tends to be linear with respect to *control* current, or the departure from quiescent current, *for a fixed displacement*, but it is also a function of displacement. For Class B operation, the restoring force tends to be a squared function of coil current *for a fixed displacement* but is also a function of displacement. Class C operation involves a no-current *dead zone* and is a complicated function of the circuitry and the electromechanical interrelations. In fact both Class A and Class B operation are modified from the simple descriptions given, by the time responses of the electrical and mechanical systems, though the superior stiffness of the active suspension usually limits the suspended body to much smaller excursions than would occur for a passive suspension of comparable frame size. Actually none of the active suspensions described in following sections are clear-cut examples of Class A, B, or C operation, and whereas these designations may be initially helpful in relating back to familiar electronic amplifier concepts, the considerations of basic importance from the standpoint of the suspension are whether it actually uses quiescent current and whether during operation to correct a displacement the coil currents are *continuous* or *discontinuous*. These concepts bring in the ideas of time sharing, multiplexing, and pulsing.

Whether or not quiescent current exists, continous operation means that when the suspended object is displaced, some coils always have currents in amounts required to produce restoring forces. Likewise whether or not quiescent current exists, discontinuous operation means that current changes required to produce restoring forces are intermittent, and dead zones may, in fact, exist during which the suspended body is free of electromagnetic forces.

Time sharing and multiplexing are somewhat related. Multiplexing is a sharing of circuits for the same or different purposes, but the sharing need not necessarily be all the time. The same circuit may be used for several different purposes but only one at a time, which is pure time sharing. Several circuits may use the same source for different purposes but in sequence. The combined use of windings for the passive magnetic suspension and the Microsyn, as illustrated in Figure 7-3, and for radial and axial suspension, as illustrated in Figures 7-13 and 7-17, are examples of pure multiplexing. The sharing of windings for force currents and signal voltages, as illustrated in Figures 9-1 and 9-2, are examples of pure multiplexing for active suspensions. None of these examples involve time sharing but accomplish considerable economy of material, space, weight, and machining time. Other examples, in subsequent sections, involve time sharing by removing the force current just long enough to use the windings to obtain a position signal without interference from the force current and to avoid reduction of the signal sensitivity by having the core material near saturation. Another mode of time sharing consists of sampling the positions of a suspended body in sequence along the respective axes and then introducing the respective corrective forces along one axis at a time. None of the time-sharing schemes mentioned is necessarily regarded as pulsing.

In pure pulsing operation, the coil currents are applied intermittently as the suspended body drifts outside a central dead zone on one side or another, the pulse shape, height, duration, and frequency being a matter of design. Numerous methods of pulsing operation have been devised. Pulsing may be combined with multiplexing. Sometimes a small amount of passive control is provided in the otherwise dead zone.

9-3 Theory of the Original Gilinson-Scoppettuolo* Suspension[3,7,8]

The action of this early suspension is explained on the basis of the single-axis operation of the star connection. The expression for centering force arising from direct currents is the same in form for a single-axis suspension as Equation (2-6):

$$f_{y'} = f_1 - f_3 = \frac{N^2 \mu A}{2} \left[\frac{I_1^2}{(g_0 - y')^2} - \frac{I_3^2}{(g_0 + y')^2} \right], \tag{9-1}$$

for small displacements along the y' axis of the four-pole suspension of Figure 9-2. It is idealized through the assumption of perfect magnetic materials and no stray capacitance. This expression can be taken to apply to small deflections along one axis of an eight-pole suspension in which the coil pairs straddle the x axis and the y axis in the usual manner. Then, for example, along the x axis, if forces and displacements are properly resolved,

$$F_{dc} = N^2 \mu A \left[\frac{(I_0 - I_c)^2}{\left(g_0 - x \cos \frac{\pi}{8}\right)^2} - \frac{(I_0 + I_c)^2}{\left(g_0 + x \cos \frac{\pi}{8}\right)^2} \right] \cos \frac{\pi}{8}, \tag{9-2}$$

in which I_0 is the quiescent current, to which control current I_c is added in one coil pair and subtracted in the diametrically opposite coil pair. For stable operation, the control current must be of the right sign and of adequate magnitude to give negative restoring force F_{dc} when displacement x is positive, and vice versa. For $x \ll g_0$ and $I_c \ll I_0$,

$$F_{dc} = \frac{N^2 \mu A}{g_0^2} \left(\frac{I_0^2 - 2I_0 I_c}{1 - 2x_n \cos \frac{\pi}{8}} - \frac{I_0^2 + 2I_0 I_c}{1 + 2x_n \cos \frac{\pi}{8}} \right) \cos \frac{\pi}{8}$$

$$\approx \frac{N^2 \mu A}{g_0^2} \left[(I_0^2 - 2I_0 I_c)\left(1 + 2x_n \cos \frac{\pi}{8}\right) - (I_0^2 + 2I_0 I_c)\left(1 - 2x_n \cos \frac{\pi}{8}\right) \right] \cos \frac{\pi}{8}$$

$$\approx - \frac{4N^2 I_0^2 \mu A}{g_0^2} \left(\frac{I_c}{I_0} \cos \frac{\pi}{8} - x_n \cos^2 \frac{\pi}{8} \right) = -K_1 I_c + K_2 x, \tag{9-3}$$

*Joseph A. Scoppettuolo, a principal engineer in the Draper Laboratory.

in which

$$K_1 = \frac{4N^2 I_0 \mu A}{g_0^2} \cos \frac{\pi}{8}$$

and

$$K_2 = \frac{4N^2 I_0^2 \mu A}{g_0^3} \cos^2 \frac{\pi}{8} = \frac{K_1 I_0}{g_0} \cos \frac{\pi}{8}.$$

The coefficient K_1 is the direct-current force sensitivity, and the coefficient K_2 is called the electromagnetic elastance, which may be regarded as a negative elastance, since the force term associated with it acts in the same direction as the displacement.

Superposed on the force due to direct current may be the force due to alternating current, Equation (7-25), which for $x \ll g_0$ is

$$F_{ac} = \frac{N^2 V^2 \mu A}{4R^2 g_0^2} \frac{4(Q^2 - Q_0 Q + 1)x_n}{(Q^2 + 1)^2} \cos^2 \frac{\pi}{8} = -K_{ac} x, \tag{9-4}$$

in which

$$K_{ac} = \frac{N^2 V^2 \mu A}{R^2 g_0^3} \frac{(Q_0 Q - Q^2 - 1)}{(Q^2 + 1)^2} \cos^2 \frac{\pi}{8},$$

or for $Q = 1$,

$$K_{ac} = \frac{N^2 V^2 \mu A}{4R^2 g_0^3} (Q_0 - 2) \cos^2 \frac{\pi}{2}.$$

This particular relation for force due to alternating current, which also is idealized, comes basically from Case (1), Equation (2-7), series tuning with voltage source, which was the original mode of passive operation. The original Gilinson-Scoppettuolo suspension therefore could be a hybrid active-passive suspension. For stable operation with respect to the force due to alternating current, K_{ac} must be positive, as has been discussed in Chapter 2. However, the hybrid suspension can be stable even though it may not be stable for the force due to alternating current acting alone. The original intent of the alternating-current circuitry was to provide position signals for control of the forces due to the direct currents, and for that purpose, K_{ac} need not necessarily be positive.

Superposition of forces due to direct and alternating currents gives

$$F_x = F_{dc} + F_{ac} = -K_1 I_c + K_2 x - K_{ac} x, \tag{9-5}$$

the result of which must be negative for stable operation. Hence even if the circuit parameters are not adjusted to give positive K_{ac}, the system can operate stably by having the $K_1 I_c$ force term dominate.

The dynamic action of the system is governed largely by the masses of the rotors, by the masses of the float and its contents, and by the damping action of the flotation fluid. If the system is subjected to a disturbing force F_d directed along the x axis, then

$$F_d + F_x - B\dot{x} = M\ddot{x},$$

or

$$F_d = M\ddot{x} + B\dot{x} + (K_{ac} - K_2)x + K_1 I_c, \tag{9-6}$$

in which M is the mass of the rotor and float system, and B is the damping coefficient of the fluid on the assumption that damping is viscous. If the relation between displacement along the x axis and the control component of direct current is as indicated in Figure 9-3,

$$I_c = K_c K_d K_p K_x x = Kx,$$

then Equation (9-6) can be written

$$F_d = M\ddot{x} + B\dot{x} + (K_{ac} + K_1 K - K_2)x. \tag{9-7}$$

When the direct and alternating currents are superposed in the suspension coils, the bridge circuit that is supposed to provide the position signal is unbalanced not only by the displacement of the rotor but also by unbalance of the direct currents, which affects the incremental permeabilities of the stator, so that the unbalance of the bridge circuit may be increased beyond the unbalance due to change in gap length. The consequence is to cause the signal

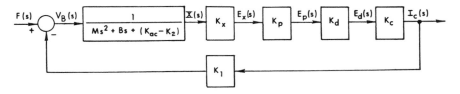

Figure 9-3 Block diagram of feedback control for system of Figure 9-2.

center to move in a direction opposite to the direction of the applied force. If as an approximation, for $x \ll g_0$ and $I_c \ll I_0$, the control signal is

$$E_{xi} \approx K_x x + K_i I_c,$$

and if the force component due to alternating current is similarly affected by the change in incremental permeability, so that

$$F_{ac} \approx -K_{ac}\left(x + \frac{K_i}{K_x}I_c\right) = -K_{ac}x - K_{xi}I_c,$$

then the conditions are as represented in Figure 9-4, which gives

$$I_c = K_c K_d K_p E_{xi},$$

or

$$I_c = \frac{K_c K_d K_p K_x}{1 - K_c K_d K_p K_i}x.$$

The force equation then becomes

$$F_d = M\ddot{x} + B\dot{x} + \left[K_{ac} + \frac{(K_1 + K_{xi})K}{1 - KK_{xi}} - K_2\right]x. \tag{9-8}$$

For the system governed either by Equation (9-7) or Equation (9-8), the amplification must be large enough to make the elastance coefficient positive, to achieve stability. In the early systems some damping was achieved by the use of lead-lag networks, and some rate feedback was introduced in the direct-current amplifier, so that the control circuitry was considerably more complex than indicated by Figure 9-3 or Figure 9-4. With little mechanical damping and substantial amplification, such circuitry is essential to limit oscillations,

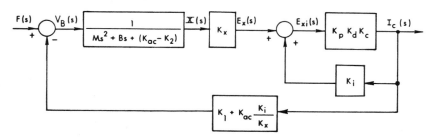

Figure 9-4 Block diagram of feedback control for system of Figure 9-2 to include the influence of change of incremental permeability.

but with the heavy fluid damping usually imposed on the float of a gyroscope, oscillating modes are unlikely to happen. Some discussion of oscillating modes and possible uses for them is given in Section 11-8. If the control current is driven very hard, so that it is not small compared with the quiescent current, the suspension can be made very stiff, but the system becomes nonlinear, the stability problem becomes more acute, and the design of the feedback control system becomes more complicated.

If in the system governed by Equation (9-8), Figure 9-4, the amplification is quite large, an anomalous situation arises. If the amplification is so small and the influence of direct current in affecting incremental permeability is so small that even with $KK_{xi} < 1$ the total elastance coefficient is negative, the system is unstable. If $KK_{xi} < 1$ and the total elastance coefficient is positive, the system is stable, with a force-displacement relation shown in Figure 9-5 as curve (a). As the amplification is increased to make $KK_{xi} \approx 1$, curve (b) is approached, which for small displacements approaches infinite stiffness. As amplification is further increased to give $KK_{xi} > 1$, curve (c) is obtained.

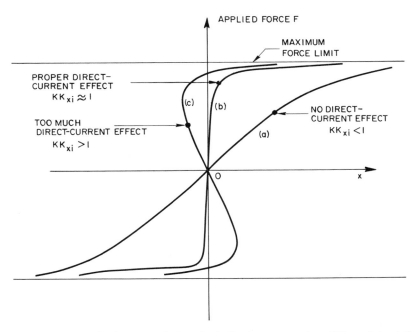

Figure 9-5 Force-displacement relations for feedback control system of Figure 9-4 as influenced by direct-current amplification.

To obtain curve (c), the disturbing force must be applied through a spring. The action may be more readily understood by visualizing it physically, rather than solely through mathematics. The force is applied through the spring slowly, so that acceleration and damping forces may be neglected. When the suspended system is given a slight positive displacement, the control signal resulting therefrom establishes a direct-current unbalance that gives a considerable restoring force reaction that promptly is further reinforced by the effective positive feedback action of the direct current in increasing the signal to call for more direct-current unbalance, so that considerable increase in force in the negative direction further deflects the pulling spring and, in fact, can give the suspended system a negative displacement. In other words, though x now is negative, the direct-current unbalance is such as would be expected to correspond to positive x. As the direct-current unbalance increases subsequent to a small positive displacement x, the restoring force brings x to zero, but the direct-current unbalance does not return to zero; this situation forces x to become negative to a balance point governed by Equation (9-8) in the steady state:

$$F_d = K_s x_s = \left[K_{ac} + \frac{(K_1 + K_{xi})K}{1 - KK_{xi}} - K_2 \right] x, \tag{9-9}$$

in which K_s and x_s are, respectively, the elastance and the elongation of the pull spring. If the pull on the spring is increased slightly, so as to decrease the negative displacement x slightly in magnitude, the direct-current unbalance is increased so as to drive x further negative to a new static balance position. The key to the action is that a small positive change in x, through large amplification and the direct-current feedback effect, generates a large increase in direct-current unbalance to drive x more negative than it was originally, because when x passes through its original position the direct-current unbalance is not thereby reduced to its original amount. The steady-state result is illustrated in Figure 9-6, in which the advance of the position of the end of the tension spring as successive points on curve (c) of Figure 9-5 are taken is designated by Δ. Thus if electromagnetic elastance

$$K_e = K_{ac} + \frac{(K_1 + K_{xi})K}{1 - KK_{xi}} - K_2,$$

then for static balance,

$$K_s(\Delta + x) = K_e x,$$

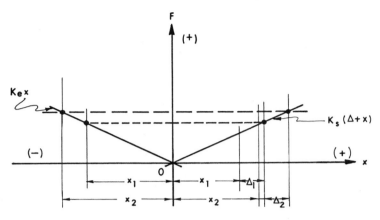

Figure 9-6 Illustration of pull-spring action in obtaining curve (c) of Figure 9-5.

or

$$x = - \frac{K_s\Delta}{K_s - K_e} = \frac{K_s\Delta}{K_e - K_s},$$ (9-10)

which must be negative for negative K_e. When pull on the spring is relaxed by returning from Δ_2 to Δ_1 of Figure 9-6, x first becomes slightly more negative and generates a large decrease in direct-current unbalance, and x is driven to a less negative balance position by the reverse of the process described. Evidently to perform this experiment with a tension spring requires $K_s < K_e$.

If the experiment is tried with small amplification, so that $KK_{xi} < 1$, but with the net electromagnetic elastance positive, a small extension Δ of the pull spring would result in a corresponding positive displacement x and in direct-current unbalance generating a force in the direction to reduce x. But now the static balance condition is

$$K_s(\Delta - x) = K_e x,$$

or

$$x = \frac{K_s\Delta}{K_s + K_e},$$ (9-11)

which for K_e positive always is positive. If now the pull spring is relaxed by returning Δ to zero, x does not go appreciably negative. With small amplification, the direct-current feedback effect is relatively small, and any tendency of x to go negative is essentially self-correcting, aside from small hysteresis effects tending to retain some alternating-current permeability unbalance. Evidently to perform this experiment with a tension spring requires $K_s > K_e$.

All of these arguments apply to small x/g_0 and hence to the lower parts of the Figure 9-5 curves. As x becomes larger, the nonlinear characteristics of magnetic materials, amplifier, demodulator, and the suspension itself give the curves considerable departures from the predictions of the simplified equations. The feedback factor K_i is not readily adjustable but is dependent on the characteristics of the magnetic materials used and the quiescent flux densities as determined by the setting of the quiescent direct current I_0. Operation usually was with KK_{xi} adjusted to be in the range from 0.25 to 0.75, thus giving a static force-displacement characteristic between curve (a) and curve (b).

9-4 The Modern Gilinson-Scoppettuolo Suspension

The principal limitation of the original Gilinson-Scoppettuolo suspension system, at least for use in missiles and space vehicles, was the size and weight of the control components and their power requirements. In 1953–1954 vacuum-tube amplifiers and demodulators, power supplies, transformers, capacitors, chokes, and wiring, even though miniaturized in accordance with the best efforts of the day, required a package about 12 inches by 4 inches by 4 inches weighing about 5 pounds and required about 15 watts for each axis of a two-axis suspension. Hence the passive alternating-current self-stabilizing suspension, which had no need for these relatively cumbersome control components, dominated that area of application for some years, even though it could not compete in stiffness and maximum force with the active suspension.

During this period of domination, several developments in devices and ideas revived an interest in the active suspension. Fortunately, about the time the United States was entering the space exploration field (circa Atlas, 1954) the transistor was making its way into electronic circuitry, along with various solid-state diodes. Presently the printed circuit board was developed. These devices permitted extreme reductions in the size, weight, and power requirements of the control circuitry for active suspensions, and the integrated circuits reduced the electrical noise that could arise from poor connections and from connecting loops and could be mounted in module form. As early as 1958, Kasparian* designed a miniaturized feedback control system for the Gilinson-Scoppettuolo suspension utilizing the latest transistors, solid-state

* Malcolm Kasparian, Jr., then staff engineer, Instrumentation Laboratory.

diodes, and miniature transformers, to give a package about $2\frac{5}{8}$ inches by $2\frac{1}{8}$ inches by $\frac{3}{4}$ inch, weighing about 5 ounces, and requiring about 0.85 watt for each axis of a two-axis suspension. As of 1970, further reductions in size, weight, and power required were achieved by Scoppettuolo to give a package about 0.93 inch by 0.45 inch by 0.44 inch, weighing less than an ounce, and requiring about 0.5 watt of quiescent power for an entire three-axis unit. Between the time of the original Gilinson-Scoppettuolo development and the time that Kasparian initiated development of the miniature control circuitry, the three-axis scheme of using two suspensions with conical surfaces had been developed, and the scheme can be used quite satisfactorily for this suspension.

In the modern Gilinson-Scoppettuolo suspension, this very small control package is applied to a somewhat smaller unit than the original suspension. with much shorter air gaps, larger polar areas, and a ferrite rotor. Even so, the stiffnesses achieved are much larger, but the maximum forces achievable are less. In the original suspension the gap reluctances dominated the magnetic circuit, whereas in the modern version the reluctances of magnetic materials tend to dominate. When the gap reluctances dominate, the design ordinarily is limited by heating; when the reluctances of the magnetic materials dominate, the design is limited by magnetic saturation. Also for the modern version the quiescent current is relatively much reduced, and aside from wiping out this small current, the operation is not push-pull, but is one-sided. The small quiescent currents tend to reduce power requirements and to reduce spurious torques. The short gaps and the ferrite rotor of the modern suspension result in operation close to the saturation limit of the magnetic material. The maximum practical flux densities in the gaps are determined by the saturation limit somewhere in the magnetic circuit, as discussed in Section 8-8. This maximum is reached when the sum of the maximum direct and alternating currents have a corresponding limit. The maximum force achievable is determined not only because of approach to the flux-density limit per se but by the accompanying loss of sensitivity of the position bridge signal caused by decrease in incremental permeability near saturation. Hence in the design and operation of this suspension, careful attention must be given to the setting of quiescent current as related to the signal current. Too low or too high a setting of the quiescent current may lead either to unstable operation or insufficiently sensitive feedback.

The maximum average pull achievable at a pole of a magnetic suspension having air-gap cross-sectional area A, fixed flux density $\mathscr{B}_{dc}$, and peak sinusoidal flux density $\mathscr{B}_{ac}$ always must be less than the pull achievable with fixed flux density and the same saturation limit $\mathscr{B}_s$:

$$\max F_{av} = \frac{\mathscr{B}_s^2 A}{2\mu}\left[1 - \frac{2\mathscr{B}_{ac}}{\mathscr{B}_s} + \frac{3}{2}\left(\frac{\mathscr{B}_{ac}}{\mathscr{B}_s}\right)^2\right].$$

The steady component of the force is reduced by the presence of the alternating component in accordance with

$$F_{dc} = \frac{\mathscr{B}_{dc}^2 A}{2\mu} = \frac{\mathscr{B}_s^2 A}{2}\left(1 - \frac{\mathscr{B}_{ac}}{\mathscr{B}_s}\right)^2.$$

The interference of the force currents and the signal currents can be avoided by having separate magnetic structures for the respective windings, as indicated in Section 9-13 and discussed more fully in Section 11-1 with some illustrations for the modern Gilinson-Scoppettuolo system.

A force-displacement curve taken along the x axis of an eight-pole suspension is shown in Figure 9-7. As is evident from this figure and force-displacement curves for other active suspensions that follow, the initial stiffness may be relatively small, or even zero, whereas for passive suspensions the initial stiffness invariably is the maximum slope of the curve, and the slope is nearly constant over the normal working range of the suspension. For an active suspension the curve generally becomes very steep in a matter of microinches of displacement, and that steep slope or high stiffness is indicative of the performance of the device rather than the initial slope. When the slope of the force-displacement curve changes considerably over the working range, analytical derivation of the curve generally is not practically possible.

Figure 9-8a shows the control circuitry for one radial axis of this system, and Figure 9-8b shows the control circuitry for the z axis. Figure 9-9 shows the force-displacement curve taken along the z axis of a pair of suspensions the x-axis displacement curve for one of which is shown in Figure 9-7. The z-axis force depends on the cone angle and the amount of control amplification for that axis. In general, with a modest cone angle insufficient for significant reduction in radial stiffness, excessive amplification is required to achieve large axial stiffness. If the z-axis restoring force must be large, requiring large coil currents at one end, the radial force capability at that end is much reduced. Likewise, if the radial force is large for one unit, the z-axis force capability in the direction of that unit is much reduced. In fact these circumstances exist for any three-axis suspension based on the principle of conical gap surfaces, whether active or passive, when forces are limited by magnetic saturation.

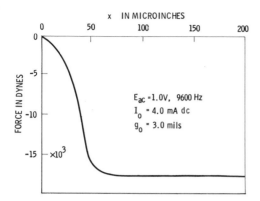

Figure 9-7 Force-displacement curve for modern Gilinson-Scoppettuolo suspension along x axis.

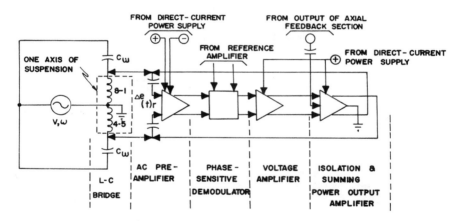

Figure 9-8a Control circuitry for modern Gilinson-Scoppettuolo suspension, x or y axis.

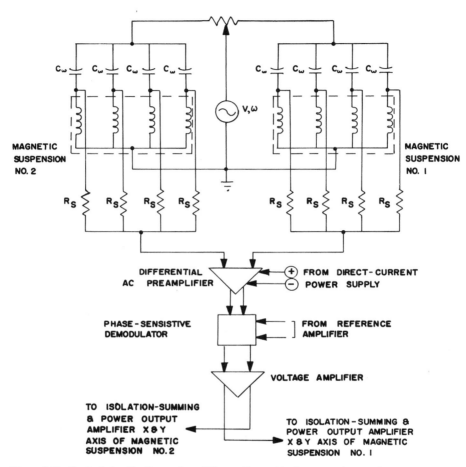

Figure 9-8b Control circuitry for modern Gilinson-Scoppettuolo suspension, *z* axis.

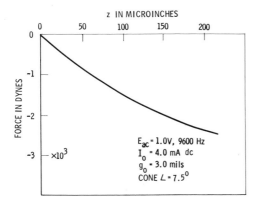

Figure 9-9 Force-displacement curve for a pair of modern Gilinson-Scoppettuolo suspensions as used for Figure 9-7 along z axis.

The positions of the cylindrical suspensions for two- or three-axis support of a gyro float are illustrated in Figure 9-10, in which the Microsyns for sensing and fixing rotational positions are separate devices. If the star connection of Figure 9-2 is applied to the coil pairs of an eight-pole suspension, then if desired it can be arranged to provide also for the functions of the Microsyn, as described in Section 7-1. In this scheme the multiplexing provides for three functions in the same windings: the force currents for radial and axial centering, and the signal currents for control of radial and axial centering, which currents serve also as the primary currents for the Microsyns that provide torque or rotational signals. But space needed for signal and torque windings might be used for suspension windings. Hence the use of multiplexing tends not only to create a magnetic saturation problem but to create a coil heating problem. But with major direct currents and minor alternating currents superposed, core losses are relatively small, so that some increase in coil heating can be afforded. Also, the combination of the magnetic suspension with the Microsyn reduces the pull area to about half the area available for a cylindrical suspension of the same size.

The quiescent positions of the two rotors of a three-axis suspension can be adjusted by potentiometers in the bridge signal circuits that move the signal center of each rotor to its desired position.

9-5 The Scoppettuolo Time-Sharing System[59]

This system is an offspring of the modern Gilinson-Scoppettuolo system. It is

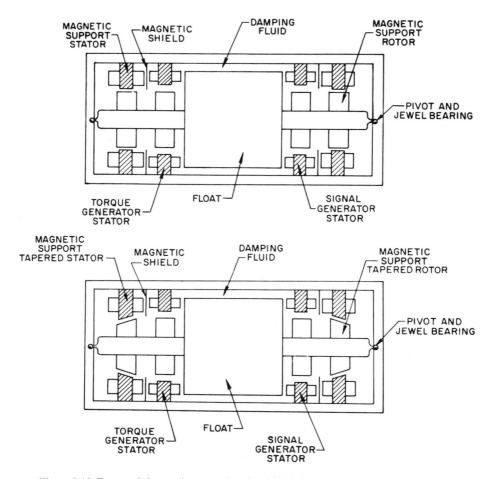

Figure 9-10 Two- and three-axis suspensions in which Microsyns are separate devices.

essentially identical, except that the direct force current is periodically re-moved while the alternating-current bridge signal is sampled. Thus the reduction of signal sensitivity by reduction of incremental impedance owing to near saturation of the magnetic circuits by the direct currents is avoided. The operation therefore is discontinuous, and while in general the average pull for such operation would be less than for continuous operation, the improved signal sensitivity achieved by periodic removal of the direct current and the somewhat higher direct flux density that can be permitted when the signal is not being sampled actually result in a substantial increase in force.

The average pull depends primarily on the length of time the direct current is applied and the length of time the suspended body is out of control, or in the *dead zone*. The ratio of the time the force is *on* to a total period, *on* time plus

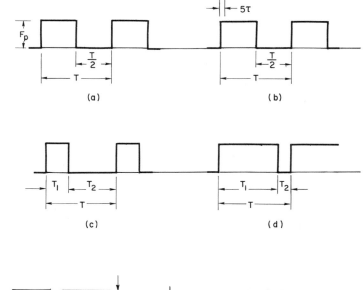

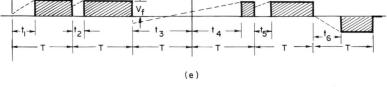

Figure 9-11 Illustrations of force pulses and duty cycles: (a) on and off times equal, (b) allowance for current buildup, (c), (d) on and off times unequal, (e) pulse initiation and duration governed by system dynamics.

off time, is called the *duty cycle*. For example, if the duty cycle is 0.50, as idealized in Figure 9-11(a), the average force is 0.50 F_p; whereas with continuous operation, full force F_p could be developed continuously, or force 0.50 F_p could be developed continuously with half the axial length of suspension. These comparisons are not exactly right, because the current-time plot cannot be perfectly rectangular; the current requires time to build up and to decay. Discontinuous operation is discussed in more detail in connection with pulsed operation. This particular time-sharing scheme is not regarded as pulsing but is regarded merely as an opening of the force-current circuit long enough to permit transients to die and obtain a relatively undisturbed sampling of the position signal. Hence the duty cycle is quite high. The scheme really involves both multiplexing and time sharing. The operation is illustrated schematically in Figure 9-12a for one radial axis of one suspension unit.

The coil pairs at opposite ends of a radial axis form a bridge circuit with auxiliary resistors R_B, which is excited continuously at 9600 hertz from a low-voltage, low-impedance source. The output of this bridge circuit is sampled at intervals 64 times the period of the excitation, or 6.7 milliseconds. The sign and amplitude of this signal are stored at the end of the sampling interval and cause direct currents to be established in the coils in directions that give centering forces. When the suspended member is centered, the coils carry quiescent currents of about 2 milliamperes. When the sampling of a bridge-circuit output indicates displacement, the direct current in the coil pair at the longer gap is increased and the current in the opposite coil pair is decreased to the point where the displacement becomes zero, whereafter it

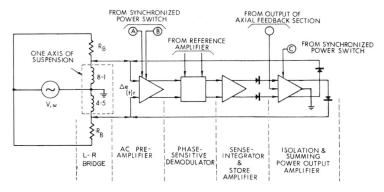

Figure 9-12a Essentials of circuitry for Scoppettuolo time-sharing system, *x* or *y* axis.

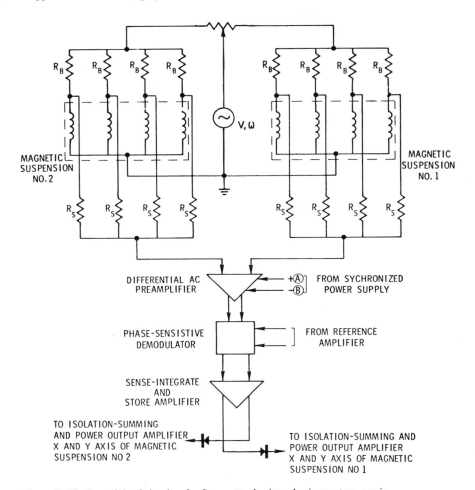

Figure 9-12b Essentials of circuitry for Scoppettuolo time-sharing system, *z* axis.

remains zero, while the current in the coil pair at the longer gap increases further as more restoring force is required. This arrangement tends to keep the quiescent power loss very low. The restoring force approximates a function proportional to the square of the direct coil current because the maximum restoring force is reached at a very small deflection. The rapidity of the sampling allows negligible wandering about zero displacement and gives essentially infinite stiffness for very small displacement, as shown in Figure 9-13(a) for radial displacement. Maximum force is reached at a displacement of several microinches. Of the total 6.7-millisecond interval $\frac{3}{16}$ is allocated to sampling, $\frac{3}{4}$ is allocated to force currents, and $\frac{1}{16}$ is allocated to decay of force currents prior to sampling, so that the nominal duty cycle is 0.75 uncorrected for rise time of the force current. The time sharing is accomplished by syn-

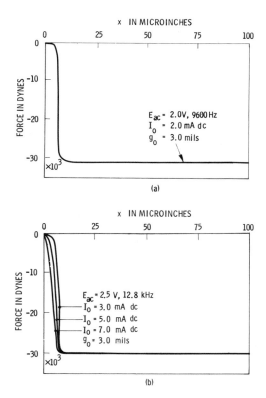

Figure 9-13 Force-displacement curves along x axis for suspension operating in Scoppettuolo time-sharing mode: (a) normal operation, (b) increased E_{ac}, f, and I_0.

chronized switching. The sampling interval of $\frac{3}{16} T$ is established by turning the preamplifier on and off through inputs A and B, Figure 9-12a, and the force-current interval of $\frac{3}{4} T$ is established by turning the output amplifier on through input C. A troublesome transient voltage spike is associated with the periodic switching of the direct force current.

The scheme can be extended to three axes by using two units having conical air-gap surfaces as previously described. The position sensing is done, as indicated in Figure 9-12b, by comparing the sum of the voltage across the four coil pairs of one unit with the corresponding summation for the other unit. For small displacements, longitudinal displacement signals are not produced by radial displacements, and vice versa. The same power amplifiers as used for the x and y axes are used for the z axis. The force-displacement curve

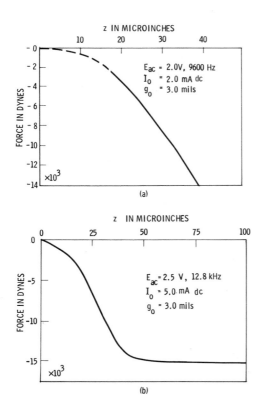

Figure 9-14 Force-displacement along z axis for a pair of suspensions as used for Figure 9-13 along x axis.

for the z axis, Figure 9-14(a), does not approach the high stiffness achieved radially, because excessive amplification would be required. For a three-axis unit, the end with the larger z-axis gaps (and hence larger radial gaps) has decreased radial stiffness because no quiescent currents and tuned circuitry exist, a result opposite to the behavior of the passive suspension, as mentioned in Section 7-6 and illustrated by Figure 7-14. These comments apply also to the Gilinson-Scoppettuolo suspensions, in contrast with the passive suspensions.

Figure 9-13(b) shows the x-axis force-displacement curves for this system for various quiescent direct currents. As the quiescent current is increased, the curves initially become steeper, and the initial square-law force-displacement characteristic disappears and changes to a nearly linear initial characteristic as should be expected. Figure 9-14(b) shows a corresponding z-axis force-displacement curve. All these force-displacement curves for the Scoppettuolo time-sharing system are for the same suspension units and differ only by having 9600 or 12,800 hertz for the alternating-current signal-bridge circuit excitation; this difference means little. They also are for the same unit used for Figures 9-7 and 9-9 for the modern Gilinson-Scoppettuolo suspension. In taking these very steep force-displacement curves, the test fixture must be extremely rigid, practically devoid of compliance. Otherwise the mechanical hysteresis of the test fixture may give loops that completely mask curves such as shown in Figure 9-13(b).

The control circuitry for this suspension differs from the circuitry for the modernized Gilinson-Scoppettuolo suspension only by the addition of circuitry for the switching to remove the direct current and sample the alternating-current signal, and hence requires a slightly larger package, about 0.93 inch by 0.52 inch by 0.44 inch for the three axes, and the power requirements are only slightly larger than required without the time sharing.

9-6 The Scoppettuolo All-Alternating-Current Active Suspension
Another means of minimizing interference of the force current with the signal current in the Gilinson-Scoppettuolo system is to substitute alternating current for the direct force current, to superpose on it signal bridge-circuit current of a much higher frequency, and to sample the bridge signal near the zero of the force current. In this system a typical bridge-circuit frequency may be 12,800 hertz and the force-current frequency may be 400 hertz. The force-current frequency is obtained by dividing down the signal-current frequency,

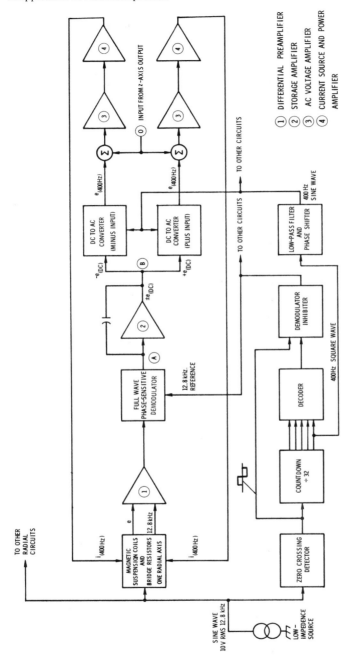

Figure 9-15 Block diagram for Scoppettuolo all-alternating current active suspension.

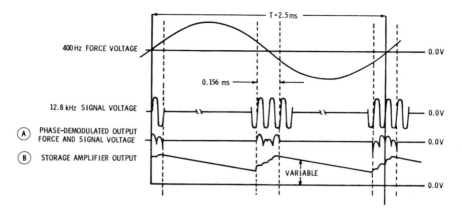

Figure 9-16 Force-voltage and signal-voltage waveforms, and sampling intervals for Scoppet-tuolo all-alternating-current active suspension.

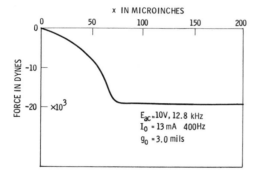

Figure 9-17 Force-displacement curve along x axis for Scoppettuolo all-alternating-current active suspension.

thus simplifying the oscillator system and also ensuring accurate phase relationship between the two frequencies. The scheme has the disadvantage that
for a given frame size, it has only about half the force capability of the direct-
current force system. Figure 9-15 shows a block diagram of the circuitry, and
Figure 9-16 shows the force and signal voltage waveforms with the sampling
interval and consequences of demodulation. Figure 9-17 shows a force-
displacement curve taken along a radial axis for the same unit as used for the
force-displacement curves for other modes of operation illustrated in this
chapter.

The Scoppettuolo time-sharing system may use alternating force current
and, by switching near zero current, avoid the troublesome transient voltage
spikes that accompany the switching of the direct current.

9-7 The Oberbeck Phase Feedback System

The Oberbeck phase feedback system, though an active system, uses alternating force current. The feedback control is continuous and the response is
essentially instantaneous, within the limitations of the supply frequency and
the time constants of the suspension-coil circuits. The action is explained
from Figure 9-18(a), which relates to a coil pair at one end of a radial axis.

Supply voltage V and voltage V_1 across the coil pair and the external resistance R_B are assumed to be sinusoidal. Then the signal voltage V_B is

$$V_B = \frac{VR_B}{R_B + 2R + 2j\omega L} = \frac{VR_B}{(R_B + 2R)\left[1 + j\left(Q'_\ell + \dfrac{Q'_0}{1 \pm x_n}\right)\right]}$$

$$\approx \frac{VR_B}{R_B + 2R} \frac{1}{\sqrt{1 + [Q'_\ell + Q'_0(1 \mp x_n)]^2}} \underline{/-\tan^{-1}[Q'_\ell + Q'_0(1 \mp x_n)]}$$

for $x_n \ll 1$, and for

$$Q'_\ell = \frac{2\omega L_\ell}{R_B + 2R},$$

$$Q'_0 = \frac{2\omega L_0}{R_B + 2R}.$$

In Block C, this signal voltage is amplified and clipped several times to give
essentially a rectangular wave shown in Figure 9-19 with amplitude V_{BC}. This
clipped voltage wave and a similar clipped reference voltage wave $V_{BC0}\underline{/90°}$,

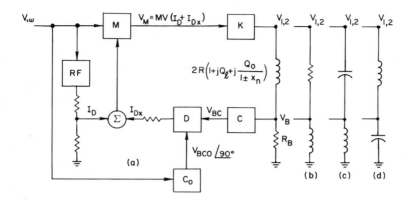

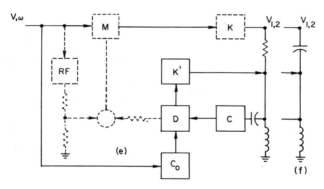

Figure 9-18 Essentials of circuitry for Oberbeck phase feedback system: (a) with LR feedback, (b) with RL feedback, (c) with CL feedback, (d) with LC feedback, (e), (f) with direct current superposed on the alternating force current.

equal to V_{BC} for $x = 0$ and shifted $90°$, are fed into demodulator D, which produces direct current

$$I_{Dx} = \pm Dx_n,$$

proportional to displacement x and of the same sign. As indicated by comparison of the V_{BC} and $V_{BC0}\underline{/90°}$ waves in Figure 9-19, their sum is zero for $x = 0$ and is positive or negative for $\pm x$. A base direct current I_D is obtained by rectification and filtering at RF and is added to I_{Dx} at modulator M. The output of this modulator is

$$V_M = MV(I_D \pm I_{Dx}),$$

and finally, by means of amplifier K,

$$V = KV_M = KMV(I_D \pm Dx_n). \tag{9-12}$$

Hence when $x = 0$, the coil pairs at the ends of the x axis are supplied with equal voltages; when displacement occurs along the x axis, the voltage applied to the coils opposite the longer gaps is increased, and the voltage applied to the coils opposite the shorter gaps is decreased. If

$$V_0 = KMVI_D$$

and

$$J = \frac{D}{I_D},$$

the voltages applied at the respective ends are

$$V_1 = V_0(1 - Jx_n) \tag{9-13}$$

and

$$V_2 = V_0(1 + Jx_n). \tag{9-14}$$

The net average restoring force, for a single-axis suspension, is

$$F_{av} = F_1 - F_2 = \frac{N^2 A}{g_0^2}\left[\frac{V_1^2}{Z_1^2(1 - x_n)^2} - \frac{V_2^2}{Z_2^2(1 + x_n)^2}\right]$$

$$= \frac{N^2 A V_0^2}{(R_B + 2R)^2 g_0^2}\left\{\frac{(1 - Jx_n)^2}{(1 - x_n)^2 + [Q_\ell'(1 - x_n) + Q_0']^2}\right.$$

$$\left. - \frac{(1 + Jx_n)^2}{(1 + x_n)^2 + [Q_\ell'(1 + x_n) + Q_0']^2}\right\}, \tag{9-15}$$

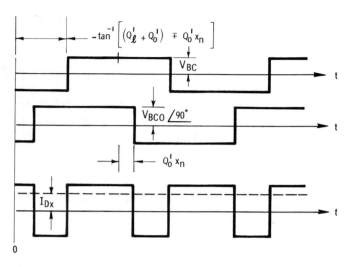

Figure 9-19 Clipped signal and reference voltages for Oberbeck phase feedback system.

in which

$$Z_1^2 = (R_B + 2R)^2 \left[1 + \left(Q'_\ell + \frac{Q'_0}{1 - x_n} \right) \right]^2$$

and

$$Z_2^2 = (R_B + 2R)^2 \left[1 + \left(Q'_\ell + \frac{Q'_0}{1 + x_n} \right) \right]^2.$$

If the leakage flux is negligible,

$$F_{av} \approx - \frac{4N^2 A V_0^2}{(R_B + 2R)^2 g_0^2} \frac{\{J[1 + (Q'_0)^2] - 1\}x_n - J(J - 1)x_n^3}{[(1 - x_n)^2 + (Q'_0)^2][(1 + x_n)^2 + (Q'_0)^2]} \qquad (9\text{-}16)$$

or

$$F_{av} \approx - \frac{4N^2 A I_0^2}{g_0^2} \frac{\{J[1 + (Q'_0)^2] - 1\}x_n - J(J - 1)x_n^3}{1 + (Q'_0)^2 + \dfrac{2[(Q'_0)^2 - 1]x_n^2 + x_n^4}{1 + (Q'_0)^2}}$$

$$\approx - \frac{4N^2 A I_0^2}{g_0^2} \frac{\left[J - \dfrac{1}{1 + (Q'_0)^2} \right]x_n - \dfrac{J(J - 1)}{1 + (Q'_0)^2}x_n^3}{1 + \dfrac{2[(Q'_0)^2 - 1]x_n^2 + x_n^4}{[1 + (Q'_0)^2]^2}}. \qquad (9\text{-}17)$$

If the force is permitted to change only at one end by placing a rectifier in series with the output of the demodulator, so that I_{Dx} always must be positive, the restoring force is

$$F_{av} = - \frac{N^2 A V_0^2}{(R_B + 2R)^2 g_0^2} \frac{\begin{array}{c} 2\{J[1 + (Q_0')^2] - 2\}x_n + J\{J[1 + (Q_0')^2] - 4\} x_n^2 \\ - 2J(J - 1)x_n^3 + J^2 x_n^4 \end{array}}{[(1 - x_n)^2 + (Q_0')^2][(1 + x_n)^2 + (Q_0')^2]}$$

(9-18)

or

$$F_{av} \approx - \frac{N^2 A I_0^2}{g_0^2} \frac{2\left[J - \dfrac{2}{1 + (Q_0')^2} \right] x_n + J\left[J - \dfrac{4}{1 + (Q_0')^2} \right] x_n^2}{1 + \dfrac{2[(Q_0')^2 - 1]x_n^2 + x_n^4}{[1 + (Q_0')^2]^2}},$$

(9-19)

approximately, if leakage flux is neglected. In Equations (9-17) and (9-19), I_0 is the rms coil current when $x = 0$.

The single-axis force expressions are developed principally to show that stable operation depends on adjustment of gain J as well as adjustment of quality ratio Q_0' for the coil circuits. For stable operation, the numerator of Equation (9-16) or Equation (9-17) must be positive when x_n is positive, that is,

$$\{J[1 + (Q_0')^2] - 1\}x_n - J(J - 1)x_n^3 > 0,$$

or

$$x_n^2 < \frac{J[1 + (Q_0')^2] - 1}{J(J - 1)}.$$

In words, unless this inequality holds up to $x_n = \pm 1$, the sign of the force reverses and the suspension becomes unstable at a displacement less than gap length g_0. A similar derivation could be based on Equation (9-18) or Equation (9-19), but with more difficulty, but since here the cubic term is the only negative term against three relatively strong positive terms, the numerator is very unlikely to be negative. In fact, for $x_n = 1$, the part of the numerator dependent on J and Q_0' becomes

$$J(Q_0')^2(J + 2) - 4, \quad .$$

which would require rather small J or Q_0' to become negative. Hence, though the use of force change at only one end considerably weakens the suspension, it decreases the likelihood of instability. These criteria are merely illustrative, being based on the equations for a single-axis suspension having no leakage flux. For an eight-pole cylindrical suspension, the projection angles that account for the radial directions of the pole pairs and the taper of the gap surfaces must be taken into account as well as the couplings among all coil pairs, so that final adjustments are best made experimentally.

A considerable convenience of this method of control is that the quiescent positioning of the suspended member does not require trimming of the working resistors or capacitors associated with the unit but can be accomplished simply by adjusting the magnitude or phase of the reference voltage V_{BCO} or the magnitude of reference current I_D. Unless the various gains in the circuits for control at opposite ends of an axis are equal, then

$$V_1 = K_1 M_1 V I_{D1} - K_1 M_1 V D_1 x_n = K_1 M_1 V I_{D1} \left(1 - \frac{D_1}{I_{D1}} x_n\right) = V_{01}(1 - J_1 x_n)$$

and

$$V_2 = K_2 M_2 V I_{D2} + K_2 M_2 V D_2 x_n = K_2 M_2 V I_{D2} \left(1 + \frac{D_2}{I_{D2}} x_n\right) = V_{02}(1 + J_2 x_n),$$

so that unless $V_{01} = V_{02}$ the quiescent position of the suspended member does not correspond to $x = 0$, and unless $J_1 = J_2$ the stiffness for positive and negative displacements are different.

The feedback can be obtained alternatively by interchange of resistance and inductance, Figure 9-18(b), or by using the usual series tuning of the passive suspension and obtaining the signal either from the voltage drops across the coils or across the capacitors, as indicated in Figure 9-18 (c and d) for the latter arrangement. Then this method can be visualized simply as a modification of the passive-suspension principle wherein the voltages impressed across the various coil pairs are made functions of the displacement of the suspended member, the voltages applied to the circuits of coil pairs that are opposite the longer gaps being increased, and vice versa. When the signal voltage is taken across a capacitor, it is

$$V_B = \frac{V X_c}{2R\sqrt{1 + (Q \pm Q_0 x_n)^2}} \Big/ -\tan^{-1}(Q \pm Q_0 x_n) - 90°,$$

and when the signal voltage is taken across a coil pair,

$$V_B = \frac{V\sqrt{1 + Q_0^2}}{\sqrt{1 + (Q \pm Q_0 x_n)^2}} \Big/ \tan^{-1} Q_0 - \tan^{-1}(Q \pm Q_0 x_n),$$

for $x_n \ll 1$. For either method of obtaining the signal,

$$F_{av} = \frac{N^2 A V_0^2}{4R^2 g_0^2} \left\{ \frac{(1 - Jx_n)^2}{(1 - x_n)^2 + [Q(1 - x_n) + Q_0 x_n]^2} \right.$$

$$\left. - \frac{(1 + Jx_n)^2}{(1 + x_n)^2 + [Q(1 + x_n) - Q_0 x_n]^2} \right\}$$

$$= - \frac{N^2 A V_0^2}{R^2 g_0^2} \frac{[(J - 1)(1 + Q^2) + Q_0 Q]x_n}{\{(1 - x_n)^2 + [Q(1 - x_n) + Q_0 x_n]^2\}}$$
$$\times \{(1 + x_n)^2 + [Q(1 + x_n) - Q_0 x_n]^2\}$$

$$(9\text{-}20)$$

and for the special case of $Q = 1$,

$$F_{av} = - \frac{2N^2 A I_0^2}{g_0^2} \frac{[2(J-1)+Q_0]x_n - [2J(J-1)-J(J-2)Q_0+JQ_0^2]x_n^3}{1+2(Q_0-1)x_n^2 + \left(\frac{Q_0^2}{2} - Q_0 + 1\right)^2 x_n^4}. \quad (9\text{-}21)$$

Of course J is not necessarily the same when the signal voltage is taken across a capacitor as when it is taken across a coil pair. For stable operation on a single-axis basis,

$$x_n^2 < \frac{(J - 1)(1 + Q^2) + Q_0 Q}{J[(J - 1)(1 + Q^2) - (J - 2)Q_0 Q + Q_0^2)]},$$

or for $Q = 1$,

$$x_n^2 < \frac{2(J - 1) + Q_0}{J[(2(J - 1) - (J - 2)Q_0 + Q_0^2)]},$$

up to $x_n = \pm 1$.

Other alternatives are indicated in Figure 9-18 (e and f), which allow for the use of direct current superposed on the alternating current through direct-current amplifier K' following demodulator D. A blocking capacitor is placed ahead of the amplifier and clipper C to prevent direct current from entering it.

By use of two tapered suspension units, a three-axis suspension can be formed without the addition of any control circuitry, the z-axis control then happens automatically as for a passively controlled suspension. Alternately, the radial-axis displacements may be controlled passively by means of tuned circuitry, and the z-axis displacements may be controlled actively merely by feeding the currents of the four circuits of a unit at one end of the z axis into a summing resistor and likewise for the unit at the other end of the z axis, the control signals being the voltage drops across the respective resistors. The control circuitry otherwise is the same as shown for active control of the radial axes.

A plot of erection time of a gyro float along the z axis for this combination of passive and active suspension is shown in Figure 9-20 compared with the erection time required when the suspension is all passive series-tuned, Case (1). For a heavily damped gyro float having clearances of only a few mils with

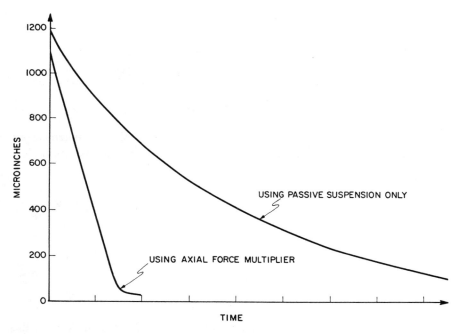

Figure 9-20 Comparison of erection times along z axis for Case (1) passive suspensions and suspensions using Oberbeck phase feedback system, Figure 9-18(a).

its casing, the time required to move it through the damping fluid from its maximum displacement to its central position along the z axis may take hours with a passive suspension, whereas the time required for radial centering may be much less. Hence the use of passive radial suspension and active axial suspension may be quite practical. The larger force of an active suspension, sustained at a higher level over the entire gap length, is a considerable advantage in shortening erection time. Though when the force current is alternating, only about half as much restoring force can be developed within the magnetic saturation limits of a given structure as for direct force current; the Oberbeck system, by having 100 percent duty cycle, avoids the loss in average force to which most direct-current systems are subject owing to pulsing or time-sharing modes of operation and delay in buildup of force currents.

When the Oberbeck phase-feedback system uses no direct force current, the problems of hysteresis effects are minimized since the rate of change of displacement is slow compared with the frequency of the supply.

9-8 Pulsed Operation

For pulsed operation a small dead zone is permitted to exist around the centered position of the suspended body, in which the body is out of control. When the suspended body is in the dead zone the suspension coils carry no force currents but may carry small signal currents. If the suspended body reaches the edge of the dead zone, a pulse of force is introduced by means of a current pulse in the direction required to drive the body back into the dead zone. The pulses may be repeated or changed in height or duration, depending on the design of the circuitry, until the displacement has been erased. Thus no quiescent component of force current is required. This procedure really involves both multiplexing and time sharing, but if the position signals can be obtained by use of the current pulses themselves, separate signal currents are not necessary, and the force current can give the maximum pull in one direction permitted by the magnetic saturation limit. However, owing to the discontinuous operation, the average pull in one direction cannot be as large as if the force were continuously applied.

A current pulse does not reach its steady state immediately but is delayed in accordance with the time constant τ of the coil circuit and may be assumed to have reached its final height after several time constants, for example, 5τ. Then, as illustrated in Figure 9-11(b), the duty cycle becomes approximately

$$\frac{1}{T}\left(\frac{T}{2} - 5\tau\right) = 0.5 - 5\frac{\tau}{T}. \tag{9-22}$$

The instantaneous force is affected somewhat also by the decay of the current, but not as much as it is affected by the current buildup, so that the average force as computed from Equation (9-22) is somewhat pessimistic, depending on the ratio τ/T. The *on* and *off* times of course need not be equal but in general could be as shown in Figure 9-11(c) or 9-11(d), but always T_1 must be long enough to permit the pulse current to build up, and T_2 must be long enough to allow it to decay, or the average force generated may be little or nothing.

The pulses indicated in Figure 9-11 (a, b, c, and d) are taken to be successive to restore the suspended member from a displaced position to its central position, or within a tolerable dead zone about its central position. The restoring pulses may be triggered by sampling a position signal at intervals, and with the suspended body adrift without control the threshold of the dead zone may be exceeded by a more or less appreciable amount before the pulse is initiated, depending on the frequency of the sampling and the speed with which the suspended body is moving. The displacement could stay within the dead zone for a considerable time, owing to heavy damping of the movable system; it could drift slowly out of the dead zone, due to forces of thermal origin; or it could change rapidly, owing to forces due to acceleration, shock, or magnetic disturbances. The duty cycle as computed applies only to consecutive pulsing periods, not to dead zone periods. If the pulses are triggered not by a position-sampling technique but immediately when a signal voltage reaches a triggering level, as illustrated by Figure 9-11(e), the action is somewhat different. If the suspended body is moving rapidly, the position signal quickly reaches the firing voltage V_f in time t_1, the force pulse is immediately initiated, and it is cut off at the end of period T. If the signal voltage still is above the firing voltage, the pulse is immediately initiated again for a full period T. If by then the displacement has been returned within the threshold, the position signal may change slowly or rapidly, so that the pulse initiation is delayed for various times t_2, t_3, ... at the outset of period T, depending on the dynamics of the system and the forces to which it is subjected. The duty cycle $(T - t - 5\tau)/T$ then is indefinite. The force and duration of one or more pulses in one direction must be sufficient to arrest the motion of the suspended body so that its displacement cannot exceed a tolerable amount, but the momentum from a single pulse must be insufficient to push the suspended body all the way through the dead zone. Hence the size of a force pulse, its duration, and the time and distance for which the suspended body may be uncontrolled are interrelated with respect to the electromechanical dynamics of the system. For this reason, pulsed operation is not applicable satisfactorily to all situations.

9-9 The Leis Pulse-Restrained Magnetic Suspension[60]

With the advent of the Apollo Task in the early 1960s, the specifications and restrictions on power, size, and weight became very severe, especially when combined with the accompanying demand for very high accuracy of performance of the guidance and navigation system. Leis reasoned that reduction of the excitation for the suspension of a floated instrument during the times when it was subjected to no external forces could be advantageous. The argument seemed attractive, because in a zero-acceleration environment, such as exists during orbital flight around a planet or its satellite, little or no force is exerted on the floated instruments.

The Leis system is a time-sharing, ternary, or three-state, scheme. In one state, the suspended member may be out of control drifting in a small central dead zone; in the other two states it is subjected to a pulse in one direction or another to return it to the dead zone if it emerges. The magnetic suspension coils are used on a time-sharing basis to serve alternately as position sensors and force coils. A simplified schematic circuit is shown in Figure 9-21 on a single-axis basis. The pulse voltage and position signal waveforms are shown in Figure 9-22, also for a single axis. In general, an eight-pole magnetic suspension is used, with two diametrically opposite pole pairs serving each axis. Auxiliary resistances R_B, Figure 9-21, are used instead of capacitors to form an alternating-current signal bridge circuit with the force-coil pairs for each axis. Use of resistors avoids the need for expensive precision capacitors. In fact resistors are necessary for a three-axis suspension to carry the force-current pulses for control along the third axis; such control does not happen automatically as for a passive suspension but requires an additional control loop, as is indicated presently. Inexpensive blocking capacitors can be used to keep direct current from the alternating-current source. The resistances R_B are so high compared with the coil resistances that very little of the steady part of a force-current pulse is diverted to the wrong coils.

At fixed intervals a sampling pulse is generated which causes the amplified error signal to be connected to a triggering circuit that, if the signal voltage has drifted beyond a predetermined threshold, releases a current pulse into the correct coil pair to force return of the suspended body toward the dead zone. Near the end of the sampling interval a reset pulse is generated that terminates the current pulse. The interval between this reset pulse and the

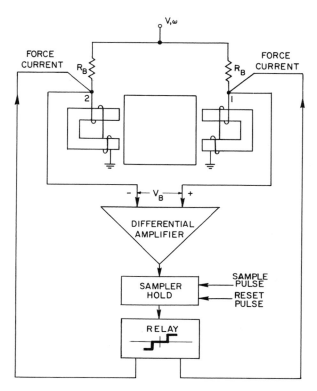

Figure 9-21 Essentials of circuitry for single-axis Leis pulse-restrained suspension.

next sampling pulse must be enough to permit the force current to decay essentially to zero and for the signal-error voltage to recover, so that the voltage between points 1 and 2, Figure 9-21, can be due to the alternating signal component only, uninfluenced by decay of the force current. If at the second sampling the signal error still is beyond the threshold, a second pulse of force current is released, and so on, until the suspended member is returned within the dead zone.

In Figure 9-22, the "error signal" is the voltage shown on Figure 9-21 between points 1 and 2. When the alternating signal voltage is between the threshold limits, the coils carry only the small alternating currents. When the alternating signal voltage is beyond one threshold or the other when sampled, it is stored as $+V$ or $-V$, regardless of the magnitude of the signal voltage, for introducing a current pulse into one coil pair or the other. When the

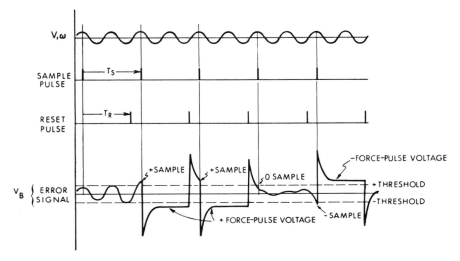

Figure 9-22 Pulse voltage and position signal voltage waveforms for Leis pulse-restrained system.

alternating signal voltage is within the thresholds, it is stored as zero, regardless of its magnitude. When a force-current pulse is terminated by a reset pulse, the $\pm V$ stored from the preceding sampling pulse is reset to zero. The amplifier is shut down with the sampling pulse and energized again with the reset pulse, so that it requires power only during the short interval between those pulses, or about 11 percent of the time. The duty cycle when the device is being pulsed is T_R/T, as indicated in Figure 9-22, about 89 percent. Actually the alternating signal current is superposed on the force-pulse current but is very small in comparison. The sign of the force-pulse voltage in Figure 9-22 has no significance with respect to the direction of the force, which depends only on the coil pair to which the current is directed.

The maximum force achievable with the Leis suspension is about $2T_R/T_S$ as much as with a passive alternating-current suspension on the same frame, series tuned, depending on Q and Q_0. For example, in comparison with the ideal single-axis, alternating-current passive suspension with $Q_0 = 10$ and $Q = 1$, for equal maximum flux densities the Leis suspension can develop a maximum force about 2.6 T_R/T_S times as much as the passive alternating-current suspension, or an average force about 2.3 times as much with 0.89 duty cycle. The advantages of the Leis suspension in this respect are that it generates no back pull and utilizes the pull corresponding to essentially maximum

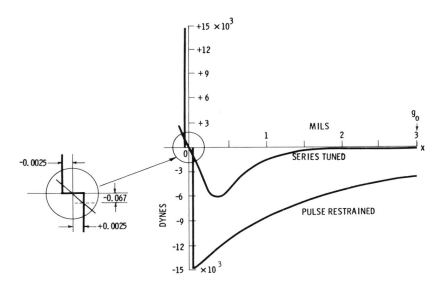

Figure 9-23 Force-displacement curves along x axis for Leis pulse-restrained suspension and for same unit operated passively, Case (1); $Q_0 = 10$, $Q = 1$.

flux density through most of the pulse. At the edge of the dead zone the Leis suspension has essentially infinite stiffness and maximum force. As the deflection increases beyond the dead-zone threshold, the force decreases, because with essentially fixed current pulses the flux density decreases. This characteristic is illustrated in Figure 9-23, in comparison with the force-displacement curve of the same unit operated passively, Case (1), with $Q_0 = 10$ and $Q = 1$.

To locate the suspended member closely, the dead zone should be narrow; to conserve power, the dead zone should be wide. A design compromise must be reached, because the momentum that may be supplied by one pulse must be insufficient to send the suspended member all the way through the dead zone from one threshold, or limit, to the other, creating a *limit cycle* or sustained state of oscillation back and forth through the dead zone.

The suspension may be extended to three axes by use of two conical members as already described and as illustrated in Figure 7-13 by use of circuitry as illustrated schematically in Figure 9-24 for the Leis scheme. Here each radial loop is as shown for the single-axis suspension, Figure 9-21. Only one loop for the x displacement (or the y displacement) of each unit is shown in Figure 9-24; a total of four such loops is required. In addition a fifth similar

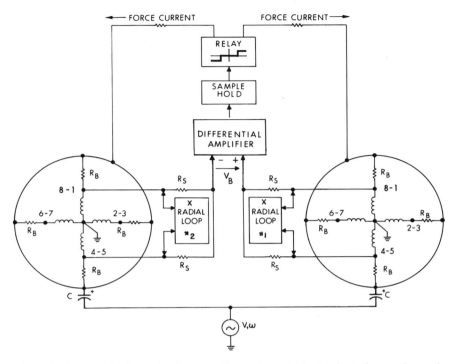

Figure 9-24 Essentials of circuitry for use of Leis pulse-restrained units to form a three-axis suspension.

loop is required for the axial or z direction. The loops for the x direction or the y direction, as shown in Figure 9-21, utilize inductances L_{81} and L_{45}, or L_{23} and L_{67} in the bridge circuits, whereas the loop for the z direction utilizes the series combination of L_{81} and L_{45} of the two units to provide the error signal through summing resistors R_s. The junction of a pair of these resistors that span L_{81} and L_{45} always is essentially at half the sum of the potentials across L_{81} and L_{45} to ground, for small displacements. When the suspended member moves in the z direction, inductances numbered 8-1 and 4-5 increase by the same amount at one end and decrease by the same amount at the other end, so that the sum of the voltage drops across them at one end is different from the sum of the drops across the corresponding inductances at the other end, and an error signal appears. When the suspended member moves in the radial direction, inductances L_{81} and L_{45} change in opposite directions by nearly the same amount, for small displacements, and essentially no z-axis error signal appears. When a z-axis error signal is beyond the dead zone at the time of sampling, a force-current pulse is sent simultaneously through all the coils and series resistances R_B at one end or the other. Transients from the force-current pulses for axial control can affect the radial centering, but theoretically the steady component of that current cannot; both transient and steady components of the force-current pulses for radial control can affect the axial centering. The sampling and reset pulses for radial and axial control are simultaneous, the interval between samplings being about 0.027 second.

The complete control system for a three-axis suspension could be packaged within a space about 1 inch by 1 inch by $\frac{1}{2}$ inch and with operation in the dead zone would require about 41.3 milliwatts total. During pulsing, the power input would increase to about 0.4 watt per axis. In contrast, a three-axis alternating-current passive suspension on a comparable frame would require about 0.64 watt, substantially independent of position, with performance comparison illustrated approximately by Figure 9-23 for one radial axis. Of the 41.3 milliwatts, 28 milliwatts are from direct-current supply, and 13.3 milliwatts are at 9600 hertz and about 0.7 lagging power factor, for bridge-circuit excitation. Thus by allowing components to be idle with little or no power input when not needed, an active suspension system having performance characteristics superior to a comparable passive suspension system was achieved without significant increase in energy or space requirements. The cost of the active system is somewhat larger than the cost of the passive system. Though the Leis system was directed toward a particular application, the principles are generally applicable.

9-10 The Hirth Pulse-Restrained Magnetic Suspension[61]

Hirth aimed his scheme at the same operating specifications used by Leis, and utilized an identical magnetic structure. His scheme also utilizes a time-sharing system wherein the same coils serve alternately to carry force currents and position-signal currents, the new features being that no alternating current is needed to produce the position signals and that the time intervals used for sensing, for application of restoring force, and for reset in readiness for the next cycle are so short that the dead zone is very narrow and the force pulses are so frequent that the force-displacement characteristic seems almost linear for small displacements. The sensing interval is 0.100 millisecond, the force-current interval is a maximum of 0.900 millisecond, and the reset interval is 0.250 millisecond, a total of 1.25 milliseconds for the cycle compared with 27 milliseconds for the approximately corresponding operations in the Leis

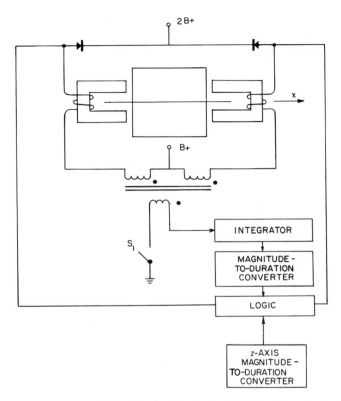

Figure 9-25 Essentials of circuitry for single-axis Hirth pulse-restrained suspension.

scheme. A schematic circuit diagram for control of one radial axis is shown in Figure 9-25 with indication of the manner of extension of the control to three axes, as explained presently.

For single-axis control, both pole pairs for that axis are energized simultaneously from $B+$ to ground in the logic box with switch S_1 closed, during the sensing interval. The consequent current pulses are in effect subtracted in the transformer, so that a difference voltage pulse is delivered by the secondary to the integrating circuit. If the suspended member is centered along the axis, the circuit inductances at the two ends are equal, the current pulses are equal, and their difference is zero. If the suspended member is displaced, the circuit inductances at the two ends are unequal, the current pulses are unequal, and a difference voltage pulse the integral of which is indicative of the magnitude and direction of the displacement results. At the end of this sensing interval the output of the integrator is transferred to a capacitor in the magnitude-to-duration converter through an amplifier. Thus at the end of each sensing interval the capacitor is charged in proportion to the accumulated or integrated voltage pulses from the transformer. When the capacitor is positively charged at the end of a sensing interval, it then is linearly discharged through current-limiting diodes; when it is negatively charged at the end of a sensing interval, it is linearly charged through current-limiting diodes. The sign and magnitude of the voltage of the capacitor are stored at the end of each sensing interval, and serve to determine which pole pair is energized by receiving force current from $B+$ to ground through the logic box and to determine the duration of the current, the limit being 0.9 millisecond. The force current is terminated by removal of the ground connection of the logic box so that the coil then discharges into $2B+$ through one of the diodes shown in Figure 9-25, which serves to reduce the current quickly during the remainder of the 1.25-millisecond period and to return most of the stored energy to the source. Thus if the suspended member is displaced from center by some steady force, for example, the average restoring force duration builds up sufficiently (if possible within 0.9 millisecond) to return the suspended member to zero and hold it there. With the suspended member centered, the

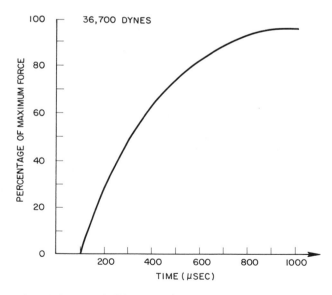

Figure 9-26 Force buildup for Hirth pulse-restrained suspension.

integral delivered to the capacitor in the magnitude-to-duration converter ceases to change. If the suspended member is forced off center in the opposite direction, the sign of the voltage pulse from the transformer changes, so that the integral first decreases and then changes sign; the restoring force likewise decreases and then reverses and builds up to an average sufficient to return the suspended member to zero and hold it there.

The duty cycle can be a maximum of 0.90/1.25, or 0.72, uncorrected for time required for force current to build up, the ultimate value being barely reached by the end of the 0.9-millisecond interval, so that the average force, as visualized from Figure 9-26, is substantially less than would be indicated by the uncorrected duty cycle, especially when the force current is terminated prior to the end of the 0.9-millisecond interval.

The control can be extended to three axes by using tapered rotors as previously described. Then two sets of control circuitry are needed at each end for the two radial axes, a total of four, and a fifth similar arrangement is needed for the z axis. For the z axis all poles at both ends are energized simultaneously for sensing, the transient responses at the two ends are compared,

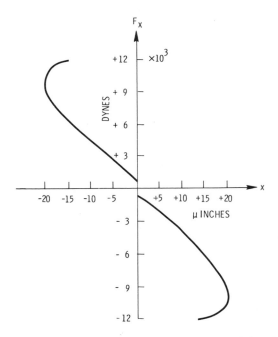

Figure 9-27 Force-displacement curve along x axis for Hirth pulse-restrained suspension.

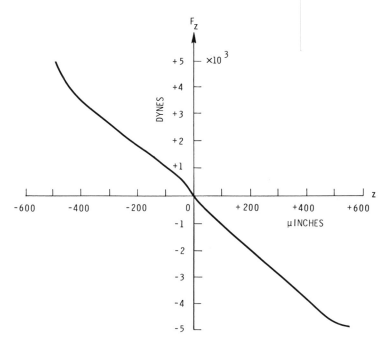

Figure 9-28 Force-displacement curve along z axis for Hirth pulse-restrained units used to form a three-axis suspension.

and all poles at one end or the other are energized simultaneously with force currents for positioning. However, the operation is such that z-axis positioning takes precedence over radial positioning. After z-axis centering has been accomplished, the remainder of a 0.9-millisecond interval (if any) can be used for radial centering; otherwise radial centering must start in the next cycle. Since these operations are not simultaneous, no interference can exist. Small radial displacements do not much affect the net inductance of the four windings of each suspension in parallel. Time sharing between longitudinal and radial axes can further reduce the duty cycles for the radial axes.

The force-displacement curves of Figures 9-27 and 9-28 could not be highly accurate owing to the shortcomings of the apparatus by means of which they were taken. Inadequate compensation in the system probably permitted some limit cycling along the radial axes, which in combination with the influence of the measuring apparatus accounts for the fuzziness near zero. Owing to pressures of time the system never has been optimized, so that the circuitry has not been shown in much detail. The system does seem to have sufficient promise to pursue its development further.

In addition to its redesign from the standpoint of stability, the system could be redesigned from the standpoint of reduction of power demand and miniaturization. As originally built the quiescent power drain is 3 watts. The interval provided for force current could be advantageously increased to about 2 milliseconds. The adequacy of the minimum interval of 0.25 millisecond for reset to ensure that the sensing pulses really start with zero coil voltage, zero coil current, and minimum residual flux is questionable. An advantage of the scheme is the absence during sensing of bias current or flux in the coils that could influence the inductance. During the sensing interval the unbalance of currents presumably gives a small net force, which should be negligible while balance is being restored. A considerable advantage of the system is its simplicity.

9-11 The Oberbeck Pulse-Restrained Magnetic Suspension

The Oberbeck system also has the merit of considerable simplicity. It uses no alternating current and no tuning or blocking capacitors. It requires only a few inexpensive resistors, a pair of silicon-controlled rectifiers, and simple voltage amplification, as shown by Figure 9-29 for one axis. The quiescent current can be made very small by use of large resistance R_c. The force currents provide the position signals, which may give substantial loss in resistance R_B, depending on the amount of amplification used.

When $v_1 = v_2$, the voltages of the control electrodes of both silicon-controlled rectifiers, $V_{b1} = V_{b2} = V_b$, are such that neither conducts. The

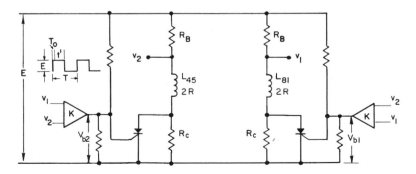

Figure 9-29 Essentials of circuitry for single-axis Oberbeck pulse-restrained suspension.

currents in the suspension coils therefore are equal and small. This situation corresponds to the suspended member centered. When the suspended member is displaced so that coils 8-1 are opposite the shorter gap, their inductance L_{81} increases, and the inductance L_{45} of coils 4-5 decreases. Then if voltage E is applied with no currents in the system,

$$i_{81} = \frac{E}{R_B + 2R + R_c}(1 - e^{-t/T_1})$$

and

$$i_{45} = \frac{E}{R_B + 2R + R_c}(1 - e^{-t/T_2}),$$

in which

$$T_1 = \frac{T_0}{1 - x_n},$$

$$T_2 = \frac{T_0}{1 + x_n},$$

and

$$T_0 = \frac{2L_0}{R_B + 2R + R_c}.$$

Since

$$v_1 = E - R_B i_1$$

and

$$v_2 = E - R_B i_2,$$

$$v_1 - v_2 = R_B(i_2 - i_1) = \frac{ER_B}{R_B + 2R + R_c}\left[e^{-(1 - x_n)t/T_0} - e^{-(1 + x_n)t/T_0}\right] \quad (9\text{-}23)$$

$$= \frac{ER_B}{R_B + 2R + R_c}e^{-t/T_0}(e^{x_n t/T_0} - e^{-x_n t/T_0}).$$

For $x_n \ll 1$,

$$v_1 - v_2 = \frac{2ER_B x_n t}{(R_B + 2R + R_c)T_0}e^{-t/T_0}. \quad (9\text{-}24)$$

When this difference has reached a maximum,

$$\frac{d(v_1 - v_2)}{dt} = 0 = \frac{2ER_Bx_n}{R_B + 2R + R_c}e^{-t/T_0}\left(\frac{1}{T_0} - \frac{t}{T_0^2}\right),$$

whence,

$$t = T_0$$

and

$$(v_1 - v_2)_{t=T_0} = \frac{0.74ER_Bx_n}{R_B + 2R + R_c}. \tag{9-25}$$

This derivation is made on the assumption that T_0 is so small that x_n cannot change significantly in that time.

Now if V_f is the firing voltage that permits a rectifier to become conducting, the rectifier associated with coil 4-5 conducts when

$$V_{b2} \geq V_f = K(v_1 - v_2) + V_b$$

if

$$V_f - V_b = K(v_1 - v_2)$$

is small compared with E, and R_c then is essentially short circuited; so current i_{45} builds up faster, from

$$i_{45} = \frac{E}{R_B + 2R + R_c}[1 - e^{-(1+x_n)}]$$

if firing occurs at $t = T_0$ to

$$i_{45} = \frac{E}{R_B + 2R}\left\{1 - \frac{[R_c + (R_B + 2R)e^{-(1+x_n)}]e^{-(1+x_n)t'/T_0'}}{R_B + 2R + R_c}\right\},$$

in which

$$T_0' = \frac{2L_0}{R_B + 2R}$$

and t' is measured from the time of short circuit of R_c. With $x_n \ll 1$ and $R_c \gg R_B + 2R$,

$$i_{45} \approx \frac{E}{R_B + 2R}(1 - e^{-t'/T_0'}). \tag{9-26}$$

For achieving a substantial duty cycle, the circuitry should be arranged to give $t' \gg T_0'$. Use of Equation (9-25) gives the threshold of the dead zone within which both rectifiers are nonconducting:

$$x_n = \frac{(R_B + 2R + R_c)(V_f - V_b)}{0.74KER_B}. \tag{9-27}$$

When $v_1 > v_2$ the silicon-controlled rectifier associated with coils 8-1 cannot fire, and current I_{81} hence remains very small. When source voltage E is reduced to zero, then since $v_1 = v_2$ again, $V_{b2} < V_f$, the silicon-controlled rectifier associated with coils 4-5 ceases to conduct, and current I_{45} becomes very small and essentially equal to current I_{81}. If the suspended body then still is displaced, it is subjected to a small net force tending to displace it further, until source voltage E is restored. If the suspended member then still is outside the dead zone, the cycle of events just described is repeated, and so on, until the suspended member is returned within the dead zone. If it crosses the other threshold, then $v_2 > v_1$, and the cycle of events as described occurs with the coil pairs and associated rectifiers and amplifiers exchanged. As mentioned in connection with the Leis scheme, the momentum of a single pulse must be insufficient to send the suspended member completely through the dead zone to avoid limit cycling. The time $T_0 + t'$ during which the source voltage E is on must be adjusted in connection with the average force to avoid this situation, and the remainder of the period T must be sufficient to permit substantial recovery of $v_1 - v_2 = 0$. The force-displacement curve for radial displacement is similar to Figure 9-23 for the Leis scheme.

The Oberbeck scheme can be extended to three axes by the use of two tapered units as previously described, but the system still is under development and the details of all the circuitry have not been finalized.

9-12 Modified Leis and Scoppettuolo Systems

Owing to the dead zone allowed by the Leis system, the suspended member may drift around somewhat without control and then at the threshold of the dead zone be subjected to a sudden and rather substantial restoring force, the jolt of which can cause troublesome transient torque and position-signal errors. To minimize these troubles, a modification suggested by Watson* can

* Howard L. Watson, an assistant director of the Draper Laboratory.

be used, whereby, within the zone which otherwise would be dead, passive control is exercised over radial and axial bandwidths of 50 to 100 micro-inches, at the extremities of which the direct-current pulsing of the Leis system takes control. This modification requires the reintroduction of tuning capacitors and the use of additional components and circuitry to manage the changeover from passive to active control, and thus somewhat increases the cost, complexity, and bulk of the control system. However in delicate applications, the resulting smoothing of operation is desirable. When the suspended member is displaced to the threshold, the direct-current pulse still exerts a considerable jolt, but the frequency of such jolts is reduced by elimination of wanderings to the threshold under the influence of small disturbing forces.

A similar smoothing also suggested by Watson can be achieved for the Scoppettuolo system by utilizing passive control from the signal-bridge circuit, with the tuning capacitors reinstated.

9-13 Summary of Methods of Signal Feedback

In the development of the passive suspension systems, various methods of obtaining position signals are shown that depend in one way or another on inductance changes caused by gap-length changes when the suspended member is displaced: unbalance of impedance-bridge circuits, differential current changes as obtained by means of summing resistors or transformers, and so on. Use of these position signals is unnecessary for the operation of the passive types of suspension; they are used only for the auxiliary purposes of initial positioning and adjusting. The active types of suspensions, however, require a position signal for feedback purposes to control the force currents that supply the restoring forces when the suspended member is displaced from its quiescent position. All of the methods of obtaining position signals described for the passive type of suspension plus others are available for use as feedback signals for the active type of suspension. A summary follows.

Inductance-capacitance bridge circuits: This method of obtaining position signals really superposes the action of a series-tuned passive suspension on the action of the active suspension. On a time-sharing basis, the times during which the alternating and direct currents may exist in the suspension coils are subject to determination in accordance with the objectives of the design.

Inductance-resistance bridge circuits: This method of obtaining position signals also superposes the force action of alternating currents that produce the signals upon the force action of the direct currents under control of the position signals, but when the suspended member is displaced the forces due to the alternating currents always tend to increase the displacement rather than reduce it. However, these destabilizing forces can be kept small, and the times during which the alternating and direct currents exist in the suspension coils is a matter of design. Bridge-circuit resistors probably are less expensive than capacitors but introduce some small additional losses into the system.

Differential transformers: Such transformers may be used to indicate the difference of currents or voltages in suspension coils and have the advantage of giving circuit isolation. If a current transformer is used, its secondary should be closed through a low impedance. Use of a voltage transformer is preferable to avoid placing a burden on the signal circuit.

Summing resistors: Use of resistors to sum currents is illustrated in Figure 9-12(b) and Figure 9-24 to give axial signals derived from a comparison of the sum of suspension coil currents in the unit at one end of the assembly with the corresponding sum for the unit at the other end of the assembly.

Phase-shift schemes: These schemes utilize the shift in phase of voltage or current in an element of an inductance-resistance or an inductance-capacitance combination as influenced by change of inductance with change of gap length when the suspended member is displaced. Examples are given in the Oberbeck phase-locked feedback system.

Comparison of time-delay effects: These methods utilize the differences in time delays or changes of currents in the suspension coils, as influenced by the changes in inductances that correspond to displaced positions of the suspended body, and require no alternating current or other auxiliary excitation. Examples are the Hirth and the Oberbeck pulse-restrained systems.

As has been suggested in Section 9-3, some velocity feedback may be desirable if mechanical damping is not heavy. In general, an alternating-current device that gives a position signal gives a velocity signal if excited by direct current or by means of a permanent magnet. Various bridge networks that involve inductances that are functions of displacement of the suspended body can function as velocity signal generators, as can various forms of auxiliary differential transformers designed for the purpose.[52]

Many variations in these schemes are conceivable, but in principle the methods summarized seem to be basic.

The position signals or the velocity signals need not be generated in the same windings or by means of the same magnetic circuits as are used for the force currents. If separate windings and magnetic structures are used, not only can interference be avoided but the designs can be separately optimized for the respective purposes. For example, for a high-frequency signal circuit, a ferrite core can be used, with appropriate adjustment of source voltage and number of turns, while for direct force currents a high-saturation-flux-density steel can be used, with appropriate adjustment of source voltage and turns. Under certain circumstances an actual reduction in power and space required may result. This possibility is discussed in Section 11-1.

9-14 The Active Electric Suspension

All the modes of operation that have been described for the magnetic suspension theoretically have their possible parallels for the electric suspension. If one side of each working capacitor must be the suspended body, then the ways in which the capacitors may be connected are limited. Likewise, requirement that the body must be grounded imposes a restriction. The parallel to magnetic saturation limit, as discussed in Section 9-4, is the dielectric breakdown limit, relating to the sum of signal and force voltages. Whereas the problem of maximum and average forces achievable relates to the maximum square and the average square of the dielectric flux density or displacement for the electric suspension, including the influence of the duty cycle when operation is discontinuous, dielectric losses enhanced by high-frequency signal voltages or high-frequency force pulsing may drastically affect dielectric breakdown, depending on the medium used. Also the effects of stray and leakage capacitances may be critical. To date the Draper Laboratory has made no serious exploration of the practicality of active electric suspensions.

9-15 Summary

The active magnetic suspension has had substantial development in recent years for use in missiles, space vehicles, and other applications in which space and energy economy are at a premium. These advances have been made possible largely through miniaturization of the control components and circuitry by use of solid-state devices and microcircuits, so that the size and weight of the control package have come down from the size of a shoe box or two, weighing ten pounds or more, to a package the size of a thumb, weighing a fraction of an ounce, and power required has been reduced from tens of watts to a fraction of a watt, for three-axis suspension of a gyro float. A secondary influence in the reduction of the size of some control components for alternating-current circuitry has been the gradual increase in operating frequency from 400 hertz to 12,800 hertz. Experimentally some systems have been operated as high as 40,000 to 50,000 hertz. The increase in operating frequency has not been sought for improvement in performance of the suspension per se. The increased operating frequency in general arises from sources available in the vehicle and intended primarily for other uses and is, in fact, detrimental to suspension performance owing to increased eddy-current and hysteresis effects, stray capacitance effects, and increased dielectric losses. Improvements in the suspension units themselves have been largely due to improvements in available magnetic materials and in machining techniques.

Control may be continuous or discontinuous and may be accomplished through auxiliary alternating-current circuitry by means of multiplexing or time sharing, or may be a by-product of the force-current circuitry itself. The force currents may be direct or alternating. A special case of discontinuous operation is pulsing, which may take a variety of forms. In fact with so many systems still under development, designs cannot be viewed as largely frozen into a few somewhat standard forms. The principal features of the active suspension in contrast with the passive suspension, however, are greater stiffness, maximum force, and power economy for comparable frame sizes. These advances have been made under the penalty of increased circuit complexity with consequent possible loss of reliability and increase in cost.

The possibilities of the active electric suspension have not been exploited.

10 TESTING AND ADJUSTING OF SUSPENSIONS

The testing of individual suspension units involves primarily the checking of the electric and magnetic circuitry, the measurement of the radial force-displacement and stiffness characteristics, and the calibration of the sensitivity of the position signals. For pairs of units assembled with a gyro float, the procedure also involves adjustment of position and measurement of axial force-displacement and stiffness characteristics as well as the calibration of the sensitivity of the axial signals. The alignment of the z axis with the gyro case is done also. The calibrated radial and axial signal sensitivities are used later when the suspension is assembled into an instrument package to check the positions of the rotors, and in particular to check the radial and axial clearances between pivots and jewels. Long-term tests are made to determine drift of rotor position due to miscellaneous small causes. Finally, indirect testing is accomplished through overall performance tests made on instrument packages of which the suspension is a component. This chapter relates primarily to the testing of the suspensions per se. The brief mention of instrument-package testing in this chapter touches merely on the principles of it in relation to component testing.

10-1 Checking of Magnetic and Electric Circuits

As discussed in Chapter 8, whereas magnetic materials need to be tested prior to being built into a suspension, they must be handled carefully in the building phase so that the operations of bonding laminations, placing of windings, encapsulation, forcing on to shafts or into housings, and final machining or grinding do not cause substantial deterioration of magnetic properties. Preliminary tests can be made to check the effects of the bonding process and final machining or grinding on individual rotors and stators against arbitrary standards by placing specified windings on them and measuring hysteresis and eddy-current losses at specified voltages and frequencies, alternating-current permeances, effects of temperature changes, and disaccommodation effects. For cores intended for active suspensions using direct force currents or pulsing, the direct-current normal permeances can be measured, and the incremental permeances at different bias levels can be measured. For units intended for operation in modern outer-space applications, tests should be made to determine the effects of nuclear radiation, if such information is not available from the manufacturer of the material or from other sources.

Such preliminary testing can eliminate cores that otherwise could result in unsatisfactory suspension assemblies. When the preliminary testing is completed, the test windings are removed, the working windings are placed and tested for short-circuited coils or grounds, and direct-current resistances are measured. This much of the testing can be done largely by standard methods.

Rotor and stator then are assembled to make a suspension unit. For most subsequent measurements special fixtures are needed, as described in following sections. The rotor position can be adjusted with respect to the stator by various means,[60] as discussed in Sections 2-8, 3-5, and 4-7. For active suspensions that operate in a small dead zone, the dead-zone position can be set by adjustment of the control signal-circuit parameters. For alternating-current suspensions, the effective coil-circuit resistances can be determined by use of a tuning capacitance C_{res} set to series resonance for the entire circuit at a desired frequency, as mentioned in Section 8-5 and illustrated in Figure 8-19. In the actual test fixture, the tuning capacitances are present, so that the rotor can be in suspension. Hence C_{res} as shown and plotted on Figure 8-19 is not the actual series resonating capacitance, but is the equivalent of the actual series capacitance for the circuit as a whole and of the four equal capacitances in series with the individual branches. The quality factors Q_{Lef} and Q can be measured. Finally the force-displacement characteristics are taken along selected axes, and tests are made for position errors, residual signals, magnetic interference, spurious signals, spurious torques, and drift. As indicated in Section 2-8, the force center and the geometric center of the suspended member do not necessarily coincide with the geometric center of the stator or with each other. Likewise the signal center does not necessarily coincide with the force center or with the geometric center of the stator. An example, for an eight-pole version of the Case (1) passive suspension is shown in Figure 10-1. Here the position signal has been taken from the inductance-capacitance bridge circuit for the x axis through a transformer and demodulated. The signal curve is slightly lopsided with respect to the force curve, due to various magnetic and geometric assymmetries.

10-2 Test Fixtures
The vertically balanced float can, Figure 10-2, was designed by Scoppettuolo[1] about 1952 and still is in use, though it has been superseded by more sophisticated apparatus for high-precision work. The float can consists of a vertical cylindrical support floated in a surrounding cylindrical can. A needle pivot bearing at the base serves principally for centering. The center of mass and

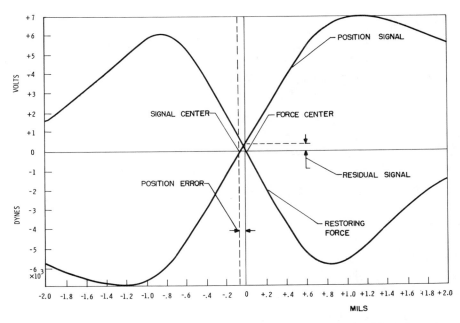

Figure 10-1 Centering force and signal voltage versus x-axis displacement for eight-pole passive suspension, Case (1) connection, illustrating shift of signal center with respect to force center.

the center of buoyancy of the floated parts are essentially coincident, so that when slightly tilted the assemblage is not subjected to appreciable net elastic torque from gravity or from flotation. Different suspension rotors under test of course may have different masses, but those masses are small compared with the total mass of the assemblage. The flotation fluid serves also to make the assemblage somewhat overdamped with respect to swinging motion about the pivot.

The distance from the pivot bearing to the component under test is sufficient that the swing motion of the test member may be regarded as planar. A four-pole cylindrical suspension structure has the coils on two opposite poles serve as two branches of an inductance-capacitance bridge circuit to give position signals along the x' axis, and the coils on the other two poles serve similarly to give position signals along the y' axis. The position signals are calibrated by micrometer measurements of the radial displacements. Alternatively a two-axis differential transformer may be used. Centering forces are measured by a delicate spring balance hooked to pull rods provided

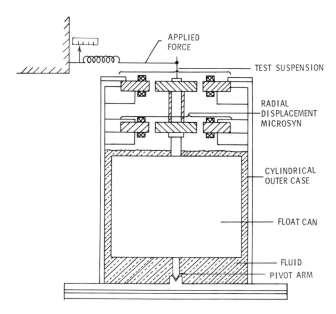

APPLIED
FORCE

TEST SUSPENSION

RADIAL
DISPLACEMENT
MICROSYN

CYLINDRICAL
OUTER CASE

FLOAT CAN

FLUID
PIVOT ARM

Figure 10-2 Vertically balanced float can.

for the purpose. Heater coils are provided to control the ambient temperature of the unit under test, if desired.

When a tapered suspension is used, the rotor must be set at exactly the right
height in the fixture to give the correct gap length. Prior to mounting in the
test can, the suspension must be assembled with the gaps gauged to the correct length. For a passive suspension using a tuning capacitor, the capacitance
then is set to give the desired quiescent power point. Then with the suspension
mounted in the test can and the rotor centered radially, its position can be
adjusted vertically until the phase angle of the input current with respect to
the input voltage is correct, whereupon the gap length should be correct. For
bridge-type passive suspensions or for active suspensions, a similar scheme
may be used merely by checking the phase angles of the coil currents when the
suspension is in the fixture against the phase angles of the coil currents when
the gaps are set.

The original pivoted-float-can method now is used mostly in production-
line testing and can be adapted to "go-no-go" procedures. It cannot be used
to obtain points beyond the peak of the force-displacement curve unless a
spring balance of large enough stiffness to stabilize the suspension and

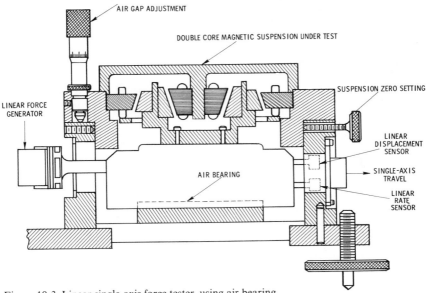

Figure 10-3 Linear single-axis force tester, using air bearing.

measuring spring system is used, and the sensitivity of such a spring generally is insufficient to make it useful as a measuring device. The float-can method is not adapted to z-axis measurements. Therefore for more general and high-precision testing an air bearing scheme has been developed, using automatic feedback control.

This system of measurement, the radial-axis version of which is illustrated in Figure 10-3, was developed by Dauwalter* and Scoppettuolo about 1959. Figure 10-4 gives an overall view. The member under test is supported by an air bearing so that it is restricted to motion only along one radial axis, essentially without friction. The member under test can be mounted to have this axis of motion along any radial direction. The displacing force is supplied by a linear voice-coil force generator, the current of which is calibrated in terms of force generated. Originally this same device carried an auxiliary coil that served to give electromagnetic viscous damping and eliminate the use of a damping fluid. Owing to undesirable coupling effects between the drive coil and the damping coil, a separate rate generator now is used for damping. Displacement is measured by a linear differential transformer of the Schaevitz

*Charles R. Dauwalter, group leader, Draper Laboratory.

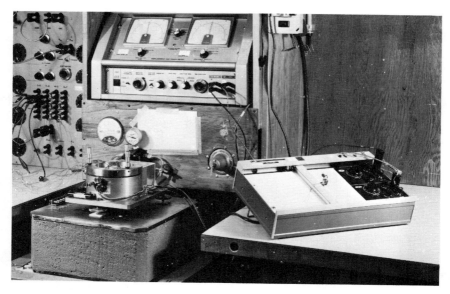

Figure 10-4 Photograph of apparatus of Figure 10-3.

type, the output of which is demodulated. A block diagram, Figure 10-5, shows the interrelations of the components. Displacements can be introduced by hand by inserting a voltage at the summing point, or automatically in the form of cyclically scheduled saw-tooth, sinusoidal, rectangular, or other waveforms. Restoring forces and displacements are plotted on an x-y recorder. Figure 10-1, for a passive suspension, was so obtained. The method can be extended to z-axis measurements. A photograph of such apparatus is shown in Figure 10-6. Here two tapered units are suspended so that motion is restricted to the z axis.

This Dauwalter-Scoppettuolo testing apparatus sometimes is called a *linear single-axis force tester* because it automatically confines motion to one direction. Unfortunately this tester cannot be used successfully for some of the new active suspensions that have extremely steep force-displacement curves, and recourse must be had to the old pivoted floated-test-can method. Particularly for such suspensions and the testing of them, not only must the mechanical compliance of the suspension assembly be made extremely small, but the mechanical compliance of the test fixture must be made practically negligible.

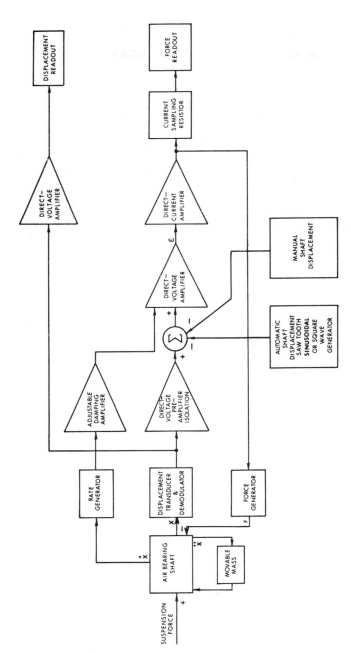

Figure 10-5 Block diagram for relations of components in apparatus of Figure 10-3.

Figure 10-6 Photograph of linear single-axis force tester having application extended to z axis.

High-precision measurements of spurious torques are made by means of a rotational air-bearing device,[63,64,65] which can detect torques on the order of 0.0001 dyne-centimeter.

10-3 Dynamic or Oscillatory Centering*

As has been indicated in connection with the use of magnetic suspensions in instruments for guidance and navigation, the suspensions are commonly used in association with Microsyns or other electromechanical devices for producing torque or angular-displacement signals about the z axis. The function of the suspensions is not only to eliminate friction about that axis but to reduce torque and signal errors that occur when rotor and stator axes are not coincident. Hence this coincidence should be as nearly perfect as possible for the quiescent condition, so that errors that occur are not chronic but occur only in accompaniment of small excursions from coincidence. Highly accurate centering is therefore very important. The use of dynamic or oscillatory centering offers a considerable improvement in accuracy over the static comparison of position signals.

* Reference 14, pp. 337–339.

The signal and torque errors of a combined Microsyn and magnetic suspension are considerably larger than the corresponding errors of a separate Microsyn unit. For example, for a separate eight-pole Microsyn unit, the angular signal error $(\Delta\theta)_r$ or the torque error $T(r)$ due to radial decentering r can be approximated by

$$\frac{(\Delta\theta)_r}{\lambda\theta_p} = \frac{T(r)}{8T_0} \approx \frac{1}{2\sqrt{2}}\left(\cos\frac{\pi}{8}\right)^2\left(\frac{r}{g_0}\right)^4 \sin 4\theta \qquad (10\text{-}1)$$

in polar coordinates, illustrated by the space plot of Figure 10-7. The angular signal or torque error of the same unit, built so that its windings serve also for the magnetic suspension, which can be approximated by

$$\frac{(\Delta\theta)_r}{\lambda\theta_p} = \frac{T(r)}{T_0} \approx \frac{1}{\sqrt{2}}(Q_c - 1)\left(\frac{Q_c}{2}\right)^4\left(\cos\frac{\pi}{8}\right)^4\left(\frac{r}{g_0}\right)^6 \sin 4\theta, \qquad (10\text{-}2)$$

is illustrated by the space plot of Figure 10-8. In these equations, θ_p is the angle subtended by a stator pole face, λ is the fraction of a stator pole face overlapped by a rotor pole face when $\theta = 0$, T_0 is the torque of one stator pole when $\theta = 0$ and the rotor is centered, and

$$\frac{r}{g_0} = \sqrt{x_n^2 + y_n^2}.$$

Inspection of Equations (10-1) and (10-2) shows that owing to $(Q_c - 1)Q_c^4$ the errors for the combined device are much larger than the errors of a separate Microsyn, except for very small radial departures. Therefore the centering of the combined device is very important, so that operation can be confined to region A, Figure 10-8, and be excluded from a region such as B.

For an example of the dynamic or oscillatory centering procedure, the error Equation (10-2) is rewritten in rectangular form:

$$\frac{(\Delta\theta)_r}{\lambda\theta_p} = \frac{T(r)}{T_0} \approx 2\sqrt{2}(Q_c - 1)\left(\frac{Q_c}{2}\right)^4\left(\cos\frac{\pi}{8}\right)^4\left(\frac{xy}{g_0^2}\right)\left(\frac{x^4 - y^4}{g_0^4}\right). \qquad (10\text{-}3)$$

If a voltage of frequency differing by 10 to 20 hertz from the excitation frequency is applied to the x-axis radial signal output terminals of a passive suspension, Figure 7-17, an oscillating radial force and corresponding radial oscillation

$$\Delta x = X \sin \omega_f t$$

occurs along the x axis. Here ω_f is the angular frequency of the force. The test generally is performed on a finished assembled instrument package as a means

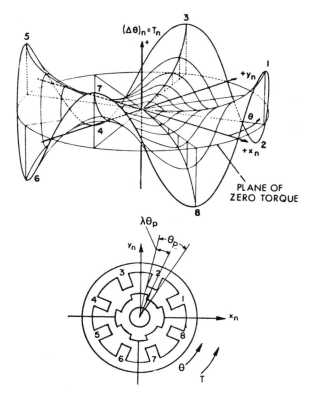

Figure 10-7 Space plot of torque errors of eight-pole Microsyn with rotor radial displacement.

of final alignment but may be performed on an individual suspension mounted in the pivoted float can. If the coordinates of the quiescent radial position are x_0, y_0, then the angular oscillation accompanying the radial oscillation is

$$\frac{(\Delta\theta)_r}{\lambda\theta_p} = 2\sqrt{2}(Q_c - 1)\left(\frac{Q_c}{2}\right)^4\left(\cos\frac{\pi}{8}\right)^4\left[\frac{(x_0 + X\sin\omega_f t)y_0}{g_0^2}\right]$$

$$\times\left[\frac{(x_0 - X\sin\omega_f t)^4 - y_0^4}{g_0^4}\right], \qquad (10\text{-}4)$$

and if by trimming the y-axis tuning capacitors the angular oscillation is stopped as evinced by cessation of the oscillation in the angular position signal of the Microsyn, then the offset y_0 must have been reduced to zero. Then

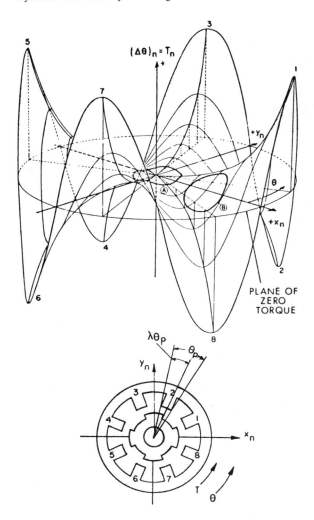

Figure 10-8 Space plot of torque errors of combined Microsyn and magnetic suspension with rotor radial displacement.

by applying the auxiliary excitation to the y-axis radial signal output terminals and trimming the x-axis tuning capacitors to stop angular oscillations the x_0 offset can be reduced to zero. The same procedure can be used with any separate Microsyn unit, with Ducosyn units, or with E-type units. The method of dynamic centering really is a general approach to the centering problem and can be adapted to the various passive suspensions and to active suspensions. For example, in the Scoppettuolo time-sharing system, Figure 9-12(a), a small alternating-voltage offset could be inserted just beyond the demodulator, which could cause oscillations along the radial axis to which that circuitry applies, with consequent angular oscillations. These angular oscillations then could be reduced to zero by adjustment of radial position along the orthogonal radial axis by trimming resistances R_B for that axis, which would indicate centering along that axis.

10-4 Float Freedom Tests
The purpose of float freedom tests is to ensure that the gyro float is free to move certain specified minimum amounts radially, axially, and rotationally. Such tests can be made by an extension of the dynamic centering procedure. For a float freedom test, the frequency of oscillation is much lower than for the dynamic centering test. Whereas too small radial and axial clearances between pivots and the jewel bearings may cause friction, too much clearance may cause difficulty in erecting the float from its rest position with the pivots on the bearings. Foreign particle contamination in the damping fluid may be detected by force "pips" in the radial or axial oscillations or by torque "pips" in the angular oscillation. The angular oscillations are obtained by applying the auxiliary excitation to the amplifier that translates the Microsyn position signal to the torquing current. This method of float freedom testing is adaptable generally to active and passive suspensions, and it and the dynamic centering procedure are readily automatically programmed, so that many prescribed combinations of float radial, axial, and angular perturbations can be obtained.

10-5 Testing for Drift of Rotor Position
Though any of the tests, adjustments, and alignment procedures mentioned or described can be made with adequate accuracy at any one time, if repeated over a period of time the results vary somewhat; the slight scattering is due not only to the precision with which the measurements can be made but also to certain changes that actually occur. This drift,[66] or uncertainty of rotor position, is caused by changes in temperature, mechanical strains, or hysteresis

due to prolonged deflections or gradual setting of encapsulating materials, magnetic disaccommodation, stray electromagnetic fields, nuclear radiation, and other environmental effects that may slightly change electric circuit parameter values, properties of magnetic materials, or other parameters that directly or indirectly affect the force balance on the suspension rotor. Especially for deep-space missions that endure for very long periods, the drift must be kept within a tolerable minimum to retain required navigational accuracy. On such missions apparatus may be dormant for long intervals to conserve energy and then may be reactivated and expected to perform satisfactorily. Therefore tests such as outlined in this chapter may be automatically programmed against a variety of environmental and operating conditions over periods of many months. Such tests may be made on complete instrument packages rather than on suspension assemblies per se to obtain the integrated effects of factors that arise from all sources.

10-6 Testing of Instrument Packages[10,67,70]

The ultimate measure of the performance of the magnetic or electric suspension comes in the test of the instrument package of which the suspension is a component. In the Draper Laboratory floated instruments intended for guidance and navigation always have been tested by relating them directly to inertial conditions. The validity of such tests has been attested by the high success of the instruments in applications on land, on sea, in the air, and in space. All the preliminary testing of materials, parts, and subassemblies, as well as the most rigid quality control in production, does not guarantee that the final package will meet rigid specifications. The testing of complete instrument packages is a varied and complicated business that must be tailored to each package, and the descriptions of the procedures are beyond the scope of this monograph. Such testing is amply described in the references cited and in other sources.

10-7 Summary

The problem of testing magnetic suspensions extends from the testing of the structural materials through the subassemblies, electric and magnetic circuitry, the final suspension assembly, and then the instrument package of which the suspension is a component. An important part of the testing of the final assembly is the careful centering of the rotor position with respect to the stator. The design of test fixtures to facilitate accurate measurements is an important engineering problem in itself.

Magnetic and electric suspensions are utilized in various instrument packages described herein, which are intended primarily for guidance and navigation purposes. Such applications include control of tanks, fixing of positions of trucks and railroad cars for mobile missile sites, fire-control systems (land based, on tanks, or on board ships), navigation of submarines and airplanes, guidance of missiles themselves and of space probes, control of space platforms, navigation of space vehicles, and mapping of fields in space. However, numerous other applications of the suspension or the principle of it are made or may be made, and various "spin-offs" have developed, some of which are discussed or mentioned in this chapter. Work continues toward increased accuracy of performance and improved testing methods and apparatus.

11-1 Separation versus Multiple Use of Magnetic and Electric Circuits
Associated with the application of the magnetic suspension to the gyro float are the need for angular displacement signals and torquing about the z axis and the need for radial and z-axis displacement signals, each of which may be provided by separate magnetic structures and separate electric circuits or by magnetic structures and electric circuits that serve two or three functions. Further, by use of tapered gaps, a pair of magnetic structures and windings may provide both axial and radial centering and, if desired, may provide angular sensing and torquing. These possibilities have been mentioned briefly in the development of the principles of operation of the passive and active suspensions.

For example, the combined Microsyn[6, 8, 52, 71, 72] and two-axis suspension is mentioned in Section 7-1 and illustrated in Figure 7-3. This combination device can provide both radial centering and angular sensing, or radial centering and torquing. Such a device has one magnetic structure, and the same windings are used for the magnetic suspension and the Microsyn primary. Section 7-1 mentions also a combination of Microsyn and two-axis magnetic suspension known as a Ducosyn,[73] illustrated in Figure 7-4, in which the magnetic and electric circuits are separate and independent, though the magnetic circuits are mounted concentrically and are coplanar. When used in pairs, the combined structure of Figure 7-3 or the suspension portion of the Ducosyn, Figure 7-4, may be tapered to give axial centering, as mentioned in Section 7-6 and illustrated in Figure 7-13. Separate Microsyn and suspension structures may be mounted coaxially and beside each other, and the units for axial centering may be independent, as mentioned in Section 7-7 and illustrated in Figure 7-19, or the suspension may be tapered, as illustrated in Figure 9-10. Strong three-axis centering may be achieved without tapering, as explained in Section 7-7 with reference to Figure 7-18(b).

Though the original development of the magnetic suspension rests considerably on the Microsyn structure and a principal objective of the centering was to reduce angular signal and torque errors of the Microsyn, this monograph does not deal with the principles of the Microsyn per se, which are covered in the references cited. However, certain interrelations between magnetic suspensions and the Microsyn are of importance here.

The object of any of these mergings of structures and windings is to economize space, but such economy always is accomplished at some sacrifice of performance; with the increasing pressures on accuracy, the trade-offs tend more and more to favor performance. Fortunately, improvements in ferromagnetic materials and more especially the development of ferrites have made possible some size and weight reduction of magnetic structures. When magnetic structures are separated, design objectives that pertain to each component can be applied separately and optimized; for example, for high-frequency signal cores a ferrite can be used, whereas for high-flux-density force cores a silicon steel can be used, and the numbers of turns and wire sizes can be set independently. For the combined Microsyn and passive magnetic suspension, because the primary magnetomotive force must serve two functions, it cannot be adjusted to serve each function best. The short gaps essential for the suspension make the signal generator or torque function unnecessarily sensitive to radial displacements and to lack of roundness of the gap surfaces. The excitation level required for the suspension, coupled with the short gaps, tends to be in excess of the optimum level for the signal generator or torquer and serves to increase signal and torque errors. Experiments show that the error torque for a suspension working on a smooth round rotor generally is less than the error torque of a suspension working on a Microsyn rotor by an order of magnitude or more. Especially for the active type of suspension, separation of cores may be advantageous, as mentioned in Section 9-13, and may actually accomplish a slight reduction in size and weight. In passive suspensions, radial and axial position signals can be obtained from the force windings with little auxiliary circuitry and with no additional burden on the windings. These signals are not needed for the operation of the suspension but are useful in centering and adjusting it. For example, the Gilinson-Scoppettuolo suspension, or the Scoppettuolo time-sharing suspension, requires position signals for feedback control, to achieve stable operation. If they are obtained from the force windings or from special coils on the same structure, either the useful force ampere-turns are limited by the presence of the signal ampere-turns, or a time-sharing scheme may be used. Either way, the use of separate units for producing force and for producing signals, each optimized for its purpose, can result in a smaller force unit and may result in a smaller overall package, and the need of sampling or time-sharing circuitry is eliminated. Further, some energy economy may result also, as a consequence of the individual designs, optimized with respect to windings and magnetic materials. A disadvantage of having separate cores and separate windings is that more leads must be brought out from the unit than would be needed if some combinations were used.

Figure 11-1 shows the unstabilizing side thrust that acts on a Ducosyn unit as a function of displacement from center position when the torque generator only is excited and the force-displacement curve of the passive magnetic suspension when it acts alone. The resultant curve shows how important close centering may be. If a very stiff active suspension, having a force-displacement curve such as illustrated in Figure 9-13(a) or 9-23 were used, the unit rarely could become sufficiently off center for the spurious force of the torque generator to become appreciable.

In Figure 11-2, the bottom curve is the force-displacement relation for a two-core active suspension of the Gilinson-Scoppettuolo type. In this device, a Ducosyn is used with the inner rotor core and direct-current inner stator windings supplying the radial force and the outer rotor core and alternating-current outer stator windings supplying the radial and angular displacement signals. The top curve shows the force-displacement relation when the alternating current for radial signals is superposed on the direct force current of the inner stator windings. The considerable reduction in force is due largely to reduction of the displacement signals caused by decrease in incremental permeability seen by the alternating currents because of the direct-current bias and partly caused by a lowering of maximum force achievable when the two currents combine to produce saturation, as explained in Section 9-1. The middle curve is illustrative of the force-displacement relation that would result with the alternating current for radial signals superposed on the direct force current of the inner stator windings, but with the single-core unit enlarged so as to have essentially the same volume as the sum of the volumes of the cores of the two-core unit. This curve is explained further presently, with reference to Figure 11-3.

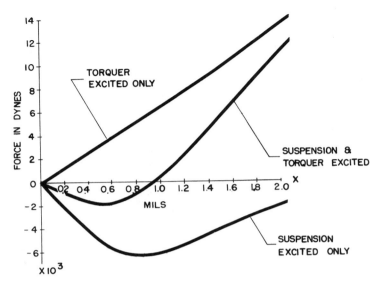

Figure 11-1 Side thrusts on Ducosyn unit due to displacement from centered position, passive suspension.

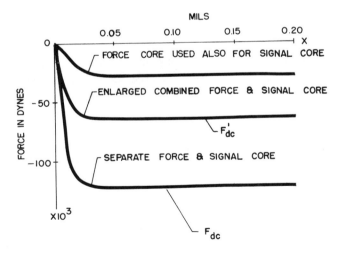

Figure 11-2 Force-displacement curves for active Gilinson-Scoppettuolo suspension having separate or combined force and signal cores.

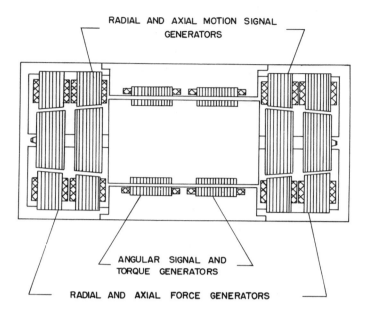

RADIAL AND AXIAL MOTION SIGNAL
GENERATORS

ANGULAR SIGNAL AND
TORQUE GENERATORS

RADIAL AND AXIAL FORCE GENERATORS

Figure 11-3 Three-axis suspension having separate angular signal and torque generators and separate radial and axial force generators.

If the cores of the Ducosyn unit are tapered, a pair of such units can also supply axial forces and axial displacement signals. If the outer rotor core surfaces are smooth, separate Microsyns are required to give angular signals or torquing. The radial and axial signal and force units may be placed beside each other instead of being coplanar, and *belly-band* Microsyn units may be used around the float to prevent the assembly from becoming so long, as illustrated by Figure 11-3. When the units that supply radial and axial signals are separate from the units that supply radial and axial forces the accurate alignment of the poles of the units is very important.

In Figure 11-3 the volume occupied by the radial and axial motion signal generators is exaggerated with respect to the volume occupied by the radial and axial force generators, the two volumes being essentially the same. Now if the middle curve of Figure 11-2 is taken to represent the combining of the functions of adjacent force and signal generators by superposing the alternating signal current and the direct force current in a winding using a core having a volume essentially equal to the sum of the volumes of the separate cores,

$$F'_{dc} = \frac{\mathscr{B}_s^2 A'}{2\mu}\left(1 - \frac{\mathscr{B}'_{ac}}{\mathscr{B}_s}\right)^2,$$

as shown in Section 9-4. Here A' represents the cross-sectional area of a polar air gap of the combined structure; if A represents the cross-sectional area of the polar air gap of the original force generator and

$$\frac{A}{A'} = 1 - \frac{\mathscr{B}'_{ac}}{\mathscr{B}_s},$$

which allocates the same total alternating flux for signal purposes as for flux density amplitude $\mathscr{B}_{ac} = \mathscr{B}_s$ in the original signal structure,

$$F'_{dc} = \frac{\mathscr{B}_s^2 A}{2\mu}\left(1 - \frac{\mathscr{B}'_{ac}}{\mathscr{B}_s}\right) = \left(1 - \frac{\mathscr{B}'_{ac}}{\mathscr{B}_s}\right)F_{dc}.$$

Evidently the combination structure must always give a smaller force than the separate structures, or, in other words, for a specified force, the separate structures should have the smaller total volume. This analysis ignores the possible deterioration of signal when the combined structure is used, due to the reduction of incremental permeability by the bias of the direct current.

Especially when very stiff active suspension units are used, the mechanical compliance of the assemblage radially may exceed the magnetic compliance of the suspension and hence cause relatively large false displacement signals. When displacements of millionths of an inch and milliseconds of arc are of concern, careful attention must be paid to the mechanical properties of the magnetic materials, wire, insulation, and potting compounds, the manner in which the parts are held together, and the manner in which the stators are anchored to their bases. Long shafts with several units strung on them are undesirable from the standpoint of bending and twisting. On the other hand, use of coplanar force and signal structures tends to give the possibility of spurious signals through the radial compliance of the force rotors and stators. These considerations also influence the question of separate versus joint use of magnetic structures. Design problems involving the compliances of the instrument package are considered in other sources.[74]

An E-type of rotational differential transformer, mentioned in Sections 7-7 and 7-8, sometimes is used instead of the Microsyn signal generator to give better separation of primary and secondary windings when the exciting frequency is high enough to make stray capacitance troublesome. The E-type torquer is not much used because it exhibits a small elastance torque. Further, the excitation frequency used for torquers generally is not high enough to require separation of control and reference coils. The E-structure may be combined with the force structure just as done in the Ducosyn. It is no better than the Microsyn with respect to giving spurious angular displacement signals or torques, due to geometric or magnetic dissymmetries or to off-center position of rotor with respect to stator. Direct-current or permanent-magnet torquers sometimes are used, having moving coils of the D'Arsonval type on a non-magnetic frame or moving iron and stationary coils. The moving-coil types are delicate and of low moment of inertia. The moving-iron types are of relatively large moment of inertia. The moving-coil torquer is used where fast rise of current and consequent fast rise of torque is required. The short rise time is made possible by the long air gaps in this type of device. However the moving-coil torquer requires flexible leads to the coil. The moving-iron type is used where short rise time is not required, because it is rugged and relatively cheap and does not have its movement restricted by flexible leads. These direct-current torquers also give spurious torques due to geometric or magnetic dissymmetries or if the rotor is off-center with respect to the stator. In general their magnetic and electric circuits are not conveniently used jointly for other functions.

11-2 A Review of the Gyro Float Erection Problem

When a magnetic suspension on a gyro float is de-energized, the float probably will drift ultimately to some radial or axial extremity with the pivots resting on the jewels. This condition may result from storage or from "shutdown" on a long voyage to conserve energy. When the time comes to activate or reactivate it, the suspension must drag the float to its central position through the surrounding fluid, or erect it. With a moderately viscous fluid and only a few mils clearance between the float and its case and in the gaps of the suspensions and Microsyns, the time required for the erection may be considerable, especially in the axial direction. To move the float axially requires that a volume of liquid essentially equal to the axial displacement of the float multiplied by the area of the end of the float must be transferred from one end of the float to the other through the thin channel between the float wall and its case. During or preliminary to the erection process, a heater may bring the fluid to a predetermined temperature and corresponding viscosity, but if the heater fails or if the thermostat fails, the suspension nevertheless should be able to erect the float under the adverse circumstances of increased fluid viscosity and buoyancy. Thus in the design of suspensions the problem of erection must be considered as well as the problems of holding position against disturbing forces and torques associated with the motions of the vehicle under guidance.

To accomplish this erection from an offset x, for example, utilizes an amount of energy

$$W = \int_x^0 F_x \, dx = \int_x^0 B_x \frac{dx}{dt} \, dx$$

for motion in the x direction. Here F_x is the restoring force of the suspension along the x axis, and B_x is the damping coefficient for velocity along the x axis. Similar equations can be written along the y and z axes. Here motion is taken to be so slow that accelerating forces are negligible. If any of the force-displacement characteristics derived are integrated, either formally or graphically, the amount of erection energy available can be thus determined. Then if coefficient B_x is measured, for example by methods described in Sections 10-3 and 10-4, the average velocity for erection can be determined, and the erection time can be computed. Evidently the area under the force-displacement curve is a measure of the erection time. To achieve erection, the suspension force of course must be directed toward the central position throughout the displacement interval.

Inspection of the various force-displacement plots for passive suspensions in Chapters 2 and 3 shows that the characteristics that give stiff centering once erection has been achieved do not give the largest energy. For example, the low-Q, high-Q_0, Case (1) or (4), plots that have large initial stiffness have sharp peaks and cover relatively little area, whereas the corresponding Case (2) or (3) plots have much smaller initial stiffness but have force peaks at much larger displacements, and hence for peaks equal to the Case (1) force peaks these plots cover much larger areas and hence give much larger erection energies. The problem is to achieve erection in a reasonable time without sacrifice of initial suspension stiffness. This objective can be accomplished to some extent by a temporary increase of excitation level during the erection period to the extent that coil heating permits and magnetic saturation makes practical. If the quiescent flux density is set too close to the saturation flux-density level, the increase in excitation may actually result in a decrease in erection energy. For example, for quiescent flux density $\mathscr{B}_0$ and control flux densities $\mathscr{B}_{c1}$ and $\mathscr{B}_{c2}$ at the respective ends of the x axis,

$$\mathscr{B}_1 = \mathscr{B}_0 - \mathscr{B}_{c1}$$

and

$$\mathscr{B}_2 = \mathscr{B}_0 + \mathscr{B}_{c2} = \mathscr{B}_s,$$

in which $\mathscr{B}_s$ represents saturation flux density. For simplicity, the action may be visualized as applying to the single-axis block suspension of Figure 2-1 with Case (1) operation. Then if

$$\mathscr{B}_{c1} = k_1 \mathscr{B}_0$$

and

$$\mathscr{B}_{c2} = k_2 \mathscr{B}_0,$$

the net restoring force, which is proportional to the difference of the squares of the flux densities at the two ends, becomes

$$F_{Vs} \propto \mathscr{B}_0^2(1 + 2k_2 + k_2^2 - 1 + 2k_1 - k_1^2) = \mathscr{B}_s^2\left[1 - \left(\frac{1 - k_1}{1 + k_2}\right)^2\right].$$

As $\mathscr{B}_0$ approaches $\mathscr{B}_s$, k_2 approaches zero; with tuning capacitance unchanged, k_1 becomes smaller. Hence as $\mathscr{B}_0$ is increased, the ratio

$$\frac{1 - k_1}{1 + k_2}$$

increases, and F_{Vs} decreases. Further temporary increase in erection energy may be had by increasing Q through an increase in tuning capacitance C, thereby shifting the force peak to a larger displacement.

Recently Oberbeck has suggested the accomplishment of this temporary shift by an increase in excitation frequency. Figure 11-4 shows the effect of a

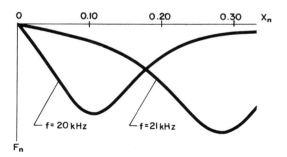

Figure 11-4 Shift of peak of centering force versus displacement for passive suspension by change of excitation frequency.

frequency increase of 5 percent on the x-axis force-displacement characteristic of a high-Q_L passive suspension connected as in Case (1). For example, if the rotor must be centered from an initial rest position of $x_n = 0.333$ but the suspension must have a high stiffness in the region $x_n < 0.1$, a large restoring force may be generated in the region of $x_n = 0.333$ by turning on the excitation at 21 kilohertz and maintaining it through a timing or feedback circuit until the restoring force falls below the 20-kilohertz curve, whereupon the excitation is shifted to 20 kilohertz.

Much of the erection trouble is avoided by use of an active suspension. As has been shown for the various active suspensions discussed in Chapter 9, not only are the stiffness and maximum force larger than achievable in passive suspensions of comparable size, but the force does not decrease so rapidly with displacement of the suspended member from centered position. Examples are in Figures 9-13(a) and 9-23. Hence for suspensions of comparable size, the areas under the force-displacement curves of active suspensions are much larger than the areas under the force-displacement curves of passive suspensions, and the erection times are correspondingly smaller. A comparison of erection times is in Figure 9-20 for erection along the z axis.

In this discussion of erection along the x axis, for example, the assumption is that displacement along the z axis is zero. If initial displacements exist both radially and axially at the same time, as ordinarily is the case, the erection problem can be much more severe, because displacement along the z axis can affect the force sensitivity along radial axes, and vice versa, either through saturation effects or through the effects of the axial shifting of tapered rotors, or a combination of the two. These adverse possibilities must be considered in the design of the suspensions. A hybrid erection scheme that eliminates the electromagnetic troubles has been suggested by Denhard.* In this scheme the float is hydrostatically centered by means of x-, y-, and z-axis orifice flow controlled by electric bridge-circuit signals. As soon as the float is within some narrow zone of displacements, the hydraulic centering ceases, and the magnetic centering of an active suspension takes over. Whereas this scheme presumably could much reduce the time constant of the erection process, it has the disadvantage of greatly adding to the cost of the instrument.

11-3 Use of Magnetic Suspensions in Fire-Control Systems

During World War II automatic fire-control systems† were developed for the

* W. G. Denhard, an associate director of the Draper Laboratory.
† Reference 14, pp. 15–31.

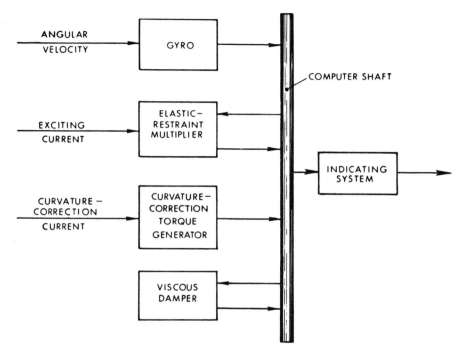

Figure 11-5 Block diagram for fire-control system computer.

directing and firing of guns on ships and aircraft and in antiaircraft installations. A block diagram of such a system is shown in Figure 11-5. The various data that enter into the directing of the gun are converted into mechanical torques and summed by the shaft to which they are applied. The gyro converts the angular velocity of the gun mount to a torque; the elastic-restraint multiplier is an adaptation of the Microsyn[71] that determines the angular position of the shaft in response to the net torque; the curvature-correction torquer also is a Microsyn that converts a current indicative of a correction[75] needed to allow for the curvature of the path of the projectile to a corresponding torque; and the viscous damper simply is a cylinder that can rotate in a viscous fluid. The angular position of the shaft indicates the position of the gun with respect to one axis and controls a servomechanism that does the positioning of the gun. This system was known as the *Draper Disturbed Line of Sight System*. In the early to middle 1950s, Draper had the idea of floating this assemblage to reduce friction and then apply magnetic suspensions to it.

Figure 11-6 Early Gilinson-Scoppettuolo active suspension control package.

Since in this system the pressure for economy of space and weight was not severe, the original Gilinson-Scoppettuolo active suspension scheme could be used very satisfactorily. The electronic control for two axes is photographed in Figure 11-6 and shows the substantial reduction in size that had been achieved by 1954.* However, most of the systems built for actual field use were fitted with passive series-tuned magnetic suspensions, owing to their relative simplicity and ruggedness. The development of new fire-control systems at the Draper Laboratory largely ceased as the Korean War came to a close. The suspended assemblage then became similar to those shown in following sections. The success of this gyro-controlled system suggested the idea of inertial guidance and navigation by means of Draper floated gyroscopes and accelerometers.

11-4 Use of Magnetic Suspensions in Guidance and Navigation Systems[76]
The guidance systems of manned ships and aircraft are usually made of special instruments monitored by human pilots. For obvious reasons automatic

*Section 9-4.

guidance equipment is required to direct the movement of missiles and un-manned spacecraft and may be desirable for manned vehicles. In the Draper inertial guidance system,[77] the governing instruments are pendulous, gyro-scopic, or a combination of such instruments. A simple block diagram, Figure 11-7, shows the control arrangement for a single axis and the location of the inertial instruments to which magnetic suspensions are applied. These instru-ments serve to govern angular directions, thrust, and integrated velocity, or distance traversed.

Navigation is primarily concerned with the determining of position and velocity of the vehicle. Guidance consists of navigation and control. Control consists of keeping the vehicle on the prescribed course in space and time. Inertial guidance is based on geometrical concepts that are essentially similar to celestial navigation but with space as determined by the fixed stars re-placed by an inertial-reference space associated with a stable member or platform, as illustrated in Figure 11-8 for control along three axes and to fix them for reference. Velocity is determined by integration of acceleration, and position is determined by a double integration of acceleration, used in con-junction with the known initial conditions. Such systems have had consider-able application in such projects as SINS, SPIRE, and SKIPPER for the guidance and navigation of submarines and other ships, planes, missiles, and space and land vehicles, and they have achieved particular success in taking the *Nautilus*, the *Skate*, and other submarines under the polar ice cap (where a magnetic compass is of little use), in the Polaris submarines and missiles, in the Atlas,[78] Titan, and Minuteman missiles, and in all of the Apollo flights.

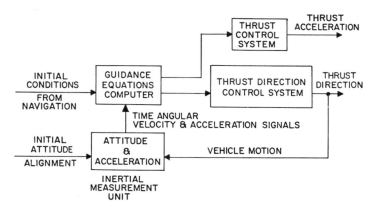

Figure 11-7 Block diagram for single-axis thrust control in Draper inertial guidance system.

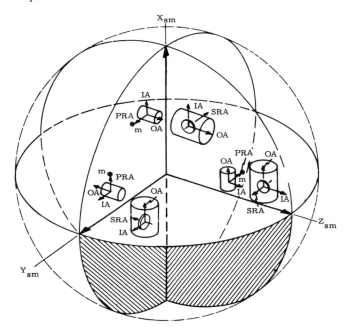

Figure 11-8 Stable platform for inertial guidance control along three axes.

The Orbiting Astronomical Observatory will be controlled by such a system. The success in the Polaris and Apollo programs[79] alone attests to the value of the magnetic suspension in the guidance and navigation system containing the Inertial Measurement Unit.

11-5 Draper Floated Instruments[12]

Five types of single-degree-of-freedom instruments were originally developed in the Draper Laboratory: (1) the Viscous Shear Integrator, (2) the Pendulous Accelerometer, (3) the Integrating Gyro, (4) the Pendulous Integrating Gyro, and (5) the Angular Velocity Integrator. Most of these instruments utilize a cylindrical float mounted concentrically within a reference cylinder.[80] A Microsyn signal generator is mounted at one end to sense the angular position of the inner cylinder with respect to the outer cylinder, and a Microsyn torquer is located at the other end to position the inner cylinder with respect to the outer cylinder in accordance with the instructions from the signal generator. The inner cylinder is immersed in a Newtonian fluid that provides the desired viscous shear and has a density to which the average density of the floated mass can be adjusted to give the desired flotation. To avoid change in density and viscosity the unit has a heater with close temperature control, the setting of which can be used for a final high-accuracy adjustment of fluid density equal to the average density of the floated parts. The floated parts then are suspended magnetically either by use of separate cylindrical suspensions, Ducosyns, or by combining the suspension and Microsyn windings.

The viscous shear integrator can be shown schematically by Figure 11-9 by removal of the pendulous mass and could serve the functions of the torque generator, viscous damper, and angle indicator shown in Figure 11-5 for the computer section of the fire-control system. In this illustration the Microsyn and suspension units make common use of the same windings and magnetic structure. The design and adjustment are such that the center of mass and the center of buoyancy of the float assembly are coincident, so that the instrument is inertially balanced in any gravitational field. The fluid supplies viscous damping laterally, longitudinally, and rotationally. The Microsyn torquer accepts an alternating-current command that develops a torque in proportion to the current amplitude. The Microsyn signal generator gives an alternating output voltage the amplitude of which is proportional to the angle

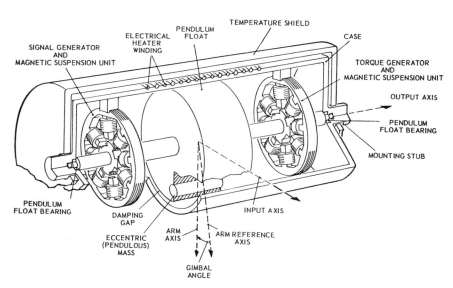

Figure 11-9 Schematic of floated pendulous accelerometer with combination Microsyn and suspension.

through which the shaft has turned against its restraining torque or is proportional to the time integral of the net torque over the interval required to achieve the final displacement. In the fire-control applications, the displacements had to be reset for each new target. With further development of the system, the integrating was done electronically.

The pendulous accelerometer has the pendulous mass as shown schematically in Figure 11-9 or schematically in Figure 11-10 and becomes a translational accelerometer. With acceleration along the input axis (IA) the pendulous mass develops a torque on the float about the output axis (OA). Any rotation about that axis is indicated by the Microsyn signal generator. Alternatively the Microsyn torquer may be fed current to develop a torque to reduce the rotation about OA to zero. Either the output voltage from the signal generator or the input current to the torquer may be taken as a measure of the acceleration along IA. Use of this instrument on a platform stable with respect to inertial space can indicate the vertical with respect to the earth, and the change in that angle is a measure of distance traversed with respect to the earth's surface by a ship or a plane. In terms of the angle between local verticals, 1 second of arc corresponds to about 90 feet on the surface of the earth. Therefore, if guidance accuracy specifications allow errors of 100 feet, accelerometers must be able to indicate angular deflections from the vertical of the order of at least 1 second of arc. Practice shows that accelerometers for inertial guidance systems should perform at least one order of magnitude better than this limit. Figure 11-11 gives a comparison of angular uncertainty of the indication of such a unit with and without a magnetic suspension.

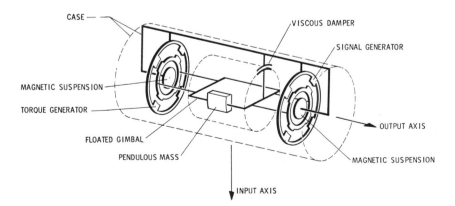

Figure 11-10 Schematic of pendulous translational accelerometer.

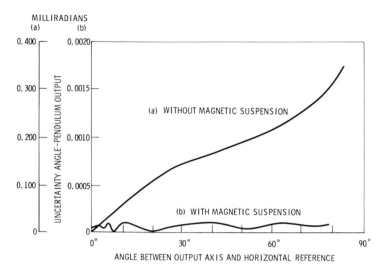

Figure 11-11 Angular uncertainty of pendulous accelerometer, with and without magnetic suspension.

Accelerations along OA tend to cause the float to rotate about IA and cause the Microsyn rotors to move radially within their stators and give small spurious angular displacement signals and torques. The desire to restrain this radial motion gave the first impetus to the development of the magnetic suspension, around 1950.

The pendulous accelerometer has considerable laboratory use as a torque-measuring device. If current is fed into the Microsyn torquer so as to rotate the pendulous mass to an equilibrium position beyond the vertical, the angle as determined by the Microsyn signal generator is a measure of the torque.

The integrating gyro[77] illustrated schematically in Figure 11-12 was developed by Draper in the 1940s. Though the gyroscope had been invented some years earlier, Draper was primarily responsible for bringing it to the state of excellence required for inertial guidance applications. It has a spinning wheel, driven at constant angular speed by a hysteresis motor, mounted on the floated gimbal frame, instead of the pendulous mass of Figure 11-10, the pendulous accelerometer. Rotation of the instrument about axis IA causes the gimbal frame to tilt about axis OA, and the change in gimbal angle is indicated by the Microsyn signal generator. Alternatively, the Microsyn torquer may be fed current to reduce the rotation about OA to zero. Since the

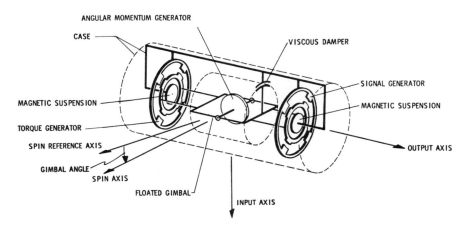

Figure 11-12 Schematic of integrating gyro.

torque developed about the axis OA by rotation of the instrument is proportional to the rate of that rotation, the angular displacement of the signal generator is proportional to the time integral of the angular velocity about IA over the interval required to achieve the final angular displacement, so that the voltage output of the signal generator or the current input to the torquer required to null the signal-generator output may be taken as a measure of angular displacement about IA, for small angles.

The pendulous integrating gyro (PIG) illustrated schematically in Figure 11-13, is a combination of the pendulous accelerometer and the integrating gyro. It was suggested by Woodbury* in the early 1950s. Here the pendulous torque is balanced by gyroscopic torque. Of itself the output of the signal generator could not distinguish between translational accelerations along IA that affect the pendulous mass, and rotations about IA that affect the gyro wheel. However the PIG is used on a platform stabilized with respect to inertial space, so that rotation about IA does not occur owing to motions of the guided vehicle, and tendency to rotate about OA is due only to acceleration along IA. To make the device operate as a translational accelerometer, additional components are required, as shown schematically in Figure 11-14, which make the entire assemblage a pendulous integrating gyro accelerometer, or PIGA.[76] The added components are a servo drive motor to rotate the PIG about IA and a resolver to indicate the angle through which IA is

* Roger B. Woodbury, a deputy director, Draper Laboratory.

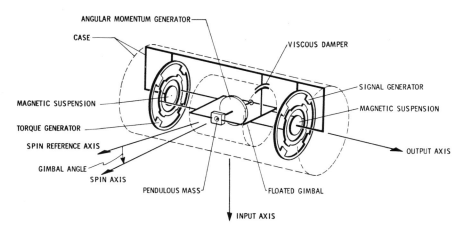

Figure 11-13 Schematic of pendulous integrating gyro (PIG).

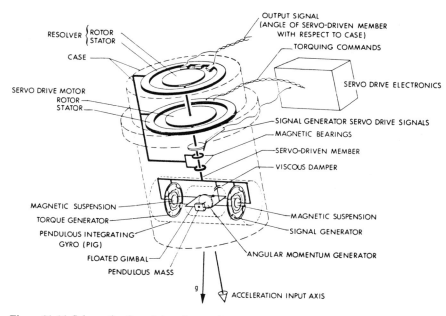

Figure 11-14 Schematic of pendulous integrating gyro accelerometer (PIGA).

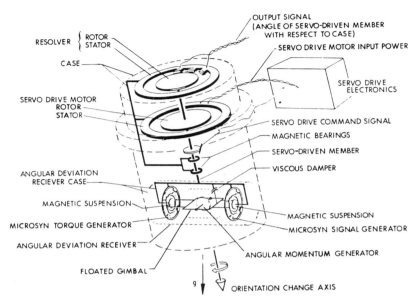

Figure 11-15 Schematic of angular velocity integrator.

rotated. The output of the Microsyn signal generator is fed to the control for the servomotor, which then is caused to drive IA of the PIG to the speed required to null the output of the signal generator by having the torque of the gyroscope about OA balance the torque of the pendulous mass about OA. The angular change of the resolver is the integral of the angular speed of IA during this adjustment interval and hence a measure of the integrated translational acceleration (translational velocity) along IA. Thus the balance is achieved by balancing two inertial torques rather than by balancing an inertial torque against an electromagnetic torque. The Microsyn torquer generally is not used as a measuring component but is left in place to retain temperature and mass symmetry in the PIG, though it may be used to establish adjusting compensations and for testing. As for the pendulous accelerometer, the accuracy of the PIG can be considerably affected by accelerations along OA which tend to cause radial deflections and accompanying signal and torque errors from the Microsyn, so that the use of magnetic suspensions for centering is highly important.

The angular velocity integrator, Figure 11-15, may be visualized by imagining the PIG of Figure 11-14 to be replaced by an integrating gyro,

Figure 11-12. If the resulting instrument is rotated about IA, a gyroscopic torque is developed about OA proportional to the speed of rotation. As for the PIGA, the output of the Microsyn signal generator is fed to the control for the servomotor, which then is caused to drive IA in the opposite direction to nullify the angle of the original disturbing rotation, and the angle is indicated by the resolver. Thus this instrument gives the same measurement as the integrating gyro alone but with greater accuracy, because the result depends on the direct matching of mechanical quantities rather than on the matching of gyro torque by electromagnetic torque. In this instrument, as in the PIGA, the Microsyn torquer, though not used as a measuring component, is retained in the system for reasons already given. In the *strapdown*[81] system (sometimes called the *hard mount* system) three of these instruments, mutually orthogonal, are fastened to the ship, plane, or other vehicle under control, and their combined outputs give the angular position of the vehicle with respect to the earth. In the Draper *base-motion-isolation* system,* three of these instruments, mutually orthogonal, are fastened to a table isolated from the vehicle by means of a gimbal system, and their outputs control servomechanisms that keep the angular position of the table unchanged with respect to inertial space, so that the angular position of the table relative to the vehicle is a reflection of the angular position of the vehicle relative to the earth.

An important use of this angular-velocity integrating principle is in the testing of gyros, for which it serves to govern a *test turntable*, or *rate table*,[†] for evaluating the drift rate of a gyro. Since the accuracy of instruments used for inertial guidance and navigation must be very high, any drift in the reference directions established by means of gyros must be kept within tolerances established in accordance with the ultimate accuracy required of the system. These tolerances are defined in terms of milli-earth-rate units (meru). The earth turns on its axis at the rate of 15 degrees per hour. Hence a meru is 0.015 degree per hour. This measuring technique can read down into the millimeru level.

11-6 Actual Instrument Packages
The illustrations in this section show the physical development of some of the actual packaged instruments, the types of which are shown schematically in the previous section. Most of these instruments were developed primarily

* Reference 12, pp. 25–26.
† Reference 12, pp. 111–123.

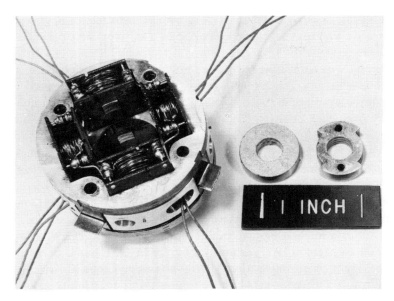

Figure 11-16 Parts of early four pole coaxial Microsyn and suspension assembly, circa 1954.

Figure 11-17 Early integrating gyro package for U.S. Navy fire-control system, circa 1954.

for space guidance or navigation, but they or similar instruments may serve on planes, ships, or ground vehicles. They span a development period of nearly 20 years.*

In Figure 11-16 are shown the parts of a four-pole Microsyn signal generator or torque generator and a four-pole magnetic suspension, circa 1954, mounted coaxially adjacent to each other. The two units are separated magnetically from each other by high-permeability shielding material. The first integrating gyro package using these units,[82] designed by Grohe† for a U.S. Navy fire-control system, had a Microsyn signal generator and a magnetic suspension at one end, a Microsyn torquer and a magnetic suspension at the other end, as shown in Figure 11-17 for size. The suspensions were mounted outboard. This package had the original Gilinson-Scoppettuolo active suspension control circuitry external to it, star connected, as shown in Figure 9-2. An early integrating gyro built at the Draper Laboratory for the Polaris guidance system in 1957 is shown in Figure 11-18. It had a spherical float with two eight-pole combination Microsyns and magnetic suspensions with conical rotors and stator gap surfaces to give z-axis constraint.

In Figure 11-19 is shown a pendulous accelerometer, circa 1957, designed by Denhard. This package has a combined Microsyn and magnetic suspension unit at each end. The suspensions in it and all following packages illustrated are passive. In Figure 11-20 is shown an integrating gyro, circa 1957, which has separate four-pole Microsyn and suspension units at each end as used in the original package, Figure 11-17, and shown in Figure 11-16 separately. The magnetic shielding can be seen in cross section between the Microsyn and suspension units, and flexible leads for the gyro motor can be seen at the left-hand end of the float. This package is approximately 6 inches in diameter by 8 inches long.

In Figure 11-21 is shown a typical pendulous integrating gyro accelerometer (PIGA) developed by Grohe for use in Polaris missiles, circa 1960. It uses untapered combined Microsyn and suspension units. The overall dimensions are about 2.5 inches in diameter by 4 inches long. In Figure 11-22 is shown an integrating gyro of about 1961 vintage developed by Hall‡ for use in a Polaris missile. It uses untapered Ducosyn units with eight-pole suspensions and Microsyns and has a spherical float. The overall dimensions are about 2.42 inches in diameter by 3.75 inches long.

* Reference 14, pp. 464–497.
† Lester R. Grohe, formerly an assistant director, Draper Laboratory.
‡ Edward J. Hall, an associate director, Draper Laboratory.

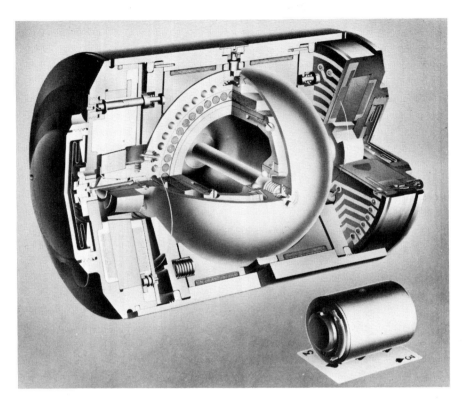

Figure 11-18 Early integrating gyro package for Polaris system, circa 1957.

In Figure 11-23 is shown an integrating gyro package developed by Hall for a Polaris application, circa 1960, and used in early Apollo guidance systems[83] to supersede the 1957 integrating gyro shown in Figure 11-18. The 1957 unit had relatively long stator poles and coils, small-diameter high-reluctance gaps, and relatively large leakage flux between poles, which reduced the magnetic stiffness. The 1960 unit had shorter stator poles and larger diameter gaps, which resulted in substantially improved performance. It used a tapered Ducosyn unit at each end with 12-pole E-type signal generator and torquer. Packages that do not use tapered suspensions depend entirely on ball-type end-thrust bearings for z-axis centering. The float is spherical. The bellows at the ends of the package allow for the expansion and contraction of the flotation fluid. The overall dimensions are about 2.42 inches in outside diameter by 3.75 inches long. Inertial guidance systems

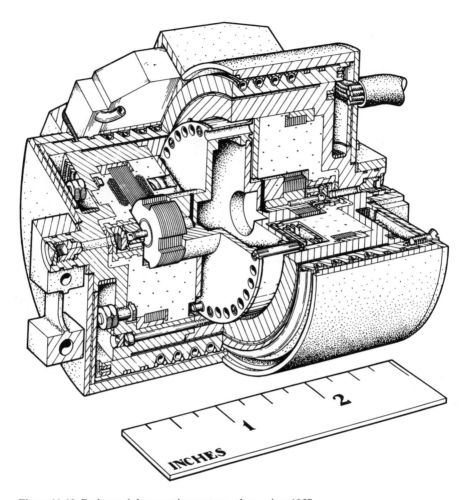

Figure 11-19 Early pendulous accelerometer package, circa 1957.

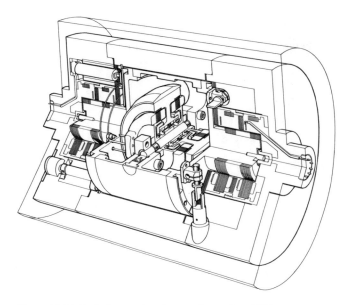

Figure 11-20 Early integrating gyro package, circa 1957.

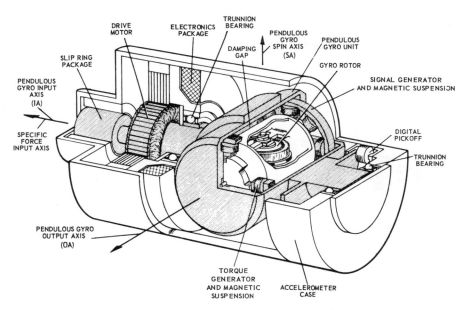

Figure 11-21 Pendulous integrating gyro accelerometer package for Polaris system, circa 1960.

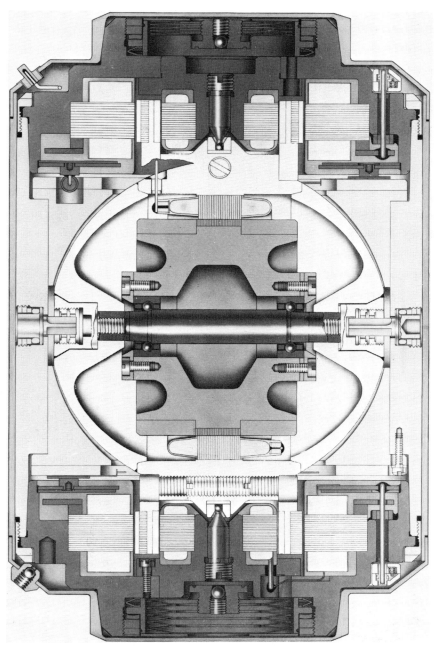

Figure 11-22 Integrating gyro package for Polaris system, circa 1961.

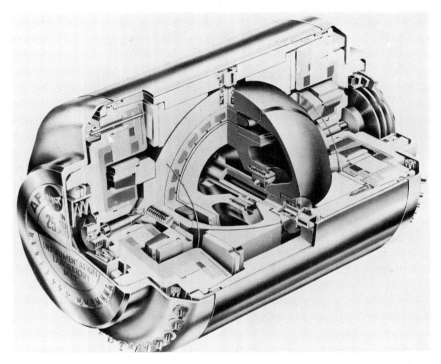

Figure 11-23 Integrating gyro package for Polaris system, circa 1960, with spherical float.

were developed also for the Poseidon program, an extension of the Polaris program, in which the missiles have longer range and carry multiple warheads. In Figure 11-24 is shown an integrating gyro package adapted by Hall for smaller Polaris guidance systems, circa 1962. It also uses a tapered Ducosyn unit at each end with a 12-pole E-type signal generator and a permanent-magnet torquer, and the float is cylindrical. The overall dimensions are about 1.98 inches in outside diameter by 3.17 inches long. In Figure 11-25 is shown an integrating gyro package circa 1966 developed by Hall especially for strapdown applications. It uses tapered Ducosyn units, the signal generator being of 12-pole E-type construction and the torquer (on the left-hand end) being of direct-current moving coil and permanent-magnet-stator construction to give fast rise of torque, as previously mentioned. Here the permanent-magnet stator is on the inside of the torquing coils. The coils around the permanent magnets are for the energizing and stabilization of the magnets after the unit is in place. The gyro wheel uses gas bearings. Gyro wheels in all units shown in previous figures use ball bearings. The overall dimensions are approximately 2.00 inches in outside diameter by 3.86 inches long.

In Figure 11-26 is shown an integrating gyro package developed by Hall about 1967 for use in a deep submergence system for a U.S. Navy underwater rescue for submarines, in which a minisubmarine searches for the disabled submarine and, upon locating it, attaches itself to it, enabling the crew of the stricken vessel to be rescued through a common hatch. This assembly uses tapered Ducosyns with a 12-pole E-type signal generator on one end and a direct-current permanent magnet torquer at the other (right-hand) end. Here the permanent-magnet stator of the torquer is outside of the torquing coils, and as for Figure 11-25 the coils around the permanent magnets are for energizing and stabilizing. The package is about 1.98 inches in diameter by 3.17 inches long.

Figure 11-27 shows a floated integrating gyro developed by Denhard that is typical of such instruments in the late 1960s. This figure shows the bellows expansion capsules that allow for expansion and contraction of the fluid. The thermal jacket and the magnetic shielding of the suspension and Microsyns against external fields are shown also. The unit is about 3 inches in outside diameter by $4\frac{1}{4}$ inches long.

A modern integrating gyro package is shown in Figure 11-28, and a modern pendulous integrating gyro accelerometer (PIGA)[84] is shown in Figure 11-29, both 1970, developed by Sapuppo,* and intended for space navigation. The

* Michele S. Sapuppo, an associate director, Draper Laboratory.

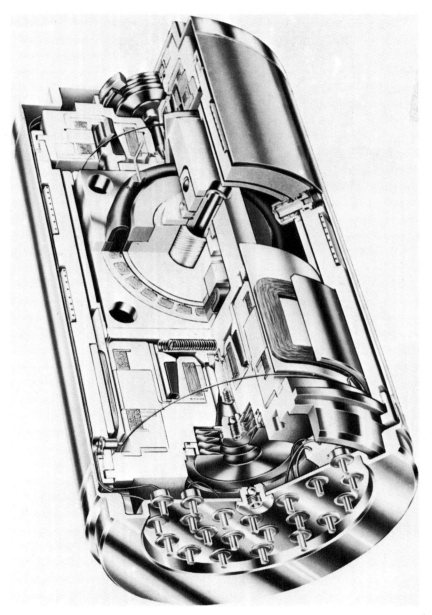

Figure 11-24 Integrating gyro package for Polaris system, circa 1962.

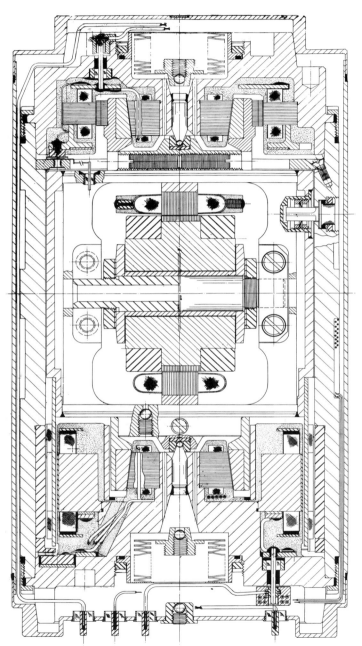

Figure 11-25 Integrating gyro package for strapdown applications, circa 1966.

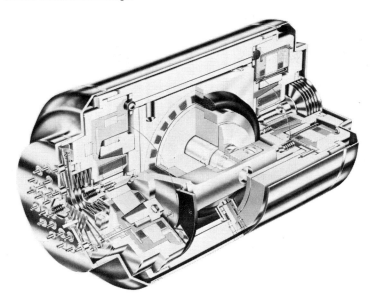

Figure 11-26 Integrating gyro package for deep submergence system for underwater rescue system, circa 1967.

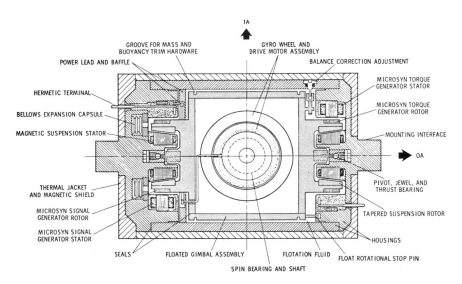

Figure 11-27 Typical integrating gyro package of late 1960s.

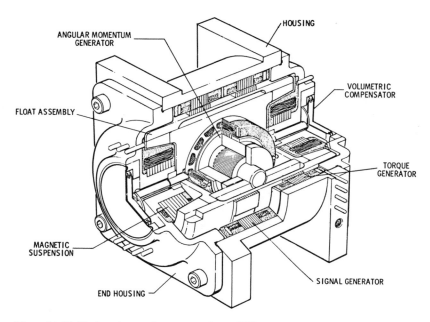

Figure 11-28 Modern integrating gyro package, 1970.

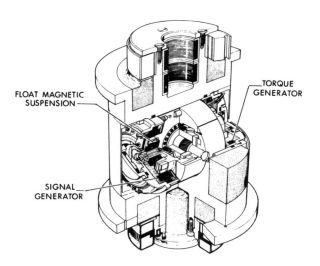

Figure 11-29 Modern pendulous integrating gyro accelerometer package, 1970.

integrating gyro has "bellyband" Microsyns, that is, the signal generator and torquer surround the float and in that position give a fatter but shorter package. The original reason for the arrangement was to fit the package into a sphere, but technical advantages are gained also: The coupling of the Microsyns and magnet suspensions is greatly reduced; the side loading of the torquer and the errors of the torquer and signal generator are considerably reduced because radial displacements caused by disturbing torques about axes normal to the float axis are considerably reduced. The PIGA uses rotary transformers to excite the signal generator on the output axis, thus avoiding the need for flexible leads. This signal generator is a *wafer resolver*, a printed-circuit device [85-87] invented by Fertig* somewhat similar to an Inductosyn. Brushes are avoided on the servomotor by means of electronic commutation. The integrating gyro is about 1.33 inches square by 1.75 inches long, and the PIGA is about 2.4 inches in diameter by 2.0 inches long.

Various other packages intended for guidance and navigation applications could be used for illustrative purposes, but the ones selected for use in this section are fairly representative of the development. The passive star-connected suspension, Figure 7-8, is used in most of the Draper floated instruments, but the advantages of other connections shown in Section 7-5 are being explored, from the standpoints of number of leads that need to be brought out and number of capacitors required, in conjunction with performance characteristics best suited to the application. Active magnetic suspensions and passive and active electric suspensions are under consideration and experimentation for use in instrument packages for which high stiffness is needed or for which low-energy requirements are paramount. Most of the Draper floated instruments use tapered stator and rotor combinations to obtain axial restraint. This method was first introduced in 1957 in the Draper 25IRIG (integrating gyro) as shown in Figure 11-18 and the Draper 25PIG (pendulous integrating gyro).[88] It has continued to be the standard method until the 1970s, but now separate axial suspensions are in favor, as explained in Section 11-1. Many up-to-date instruments developed for the armed services are presently classified and cannot be described here.

11-7 Suspension Oscillations; Dynamic Stability

A suspension may have *forced* oscillations, *natural* oscillations, or *self-sustained* oscillations that may be viewed as a type of dynamic instability. For

---

*Kenneth Fertig, chief scientist, Draper Laboratory.

a passive suspension, forced oscillations may arise from the alternating exciting currents, which give alternating forces having double the frequency of the excitation. Ordinarily, when suspensions are applied to a floated instrument, the fluid damping is so heavy that response to the double-frequency forces is inappreciable. For either active or passive suspensions that support a float containing a gyro wheel, forced oscillations may arise from unbalance of the wheel. The frequency of such oscillations coincides with the angular speed of the wheel. For a wheel driven by a four-pole hysteresis motor, for example, from a 400-hertz source, the angular speed is 200 hertz, which may be low enough to permit appreciable response. The float then oscillates in an angular motion about the spin axis SA, and the suspension rotors on the float shaft consequently move in circular arcs in the plane of the input axis IA and the output axis OA. These axes are identified in Figures 11-12 and 11-13. Active or passive suspensions may have natural oscillations that arise from external disturbing forces or from changes in the suspension forces. Depending on the electromechanical parameters of the system, the motion of the suspended member may be overdamped or underdamped. For a floated instrument the motion ordinarily is highly overdamped, so that a displacement is not oscillatory, but if the suspended member is underdamped, oscillatory motion occurs at the natural frequency of the system. For an active suspension, self-sustained oscillations may occur in the absence of sufficient damping, which are a consequence of improper design of the feedback-control circuitry, and in accordance with conventional servomechanisms theory are regarded as unstable operation. These oscillations do not represent an instability in the same sense as the static instability discussed in Section 2-4, in that they do not grow indefinitely in amplitude but settle to a repeated oscillation of fixed amplitude and frequency. Similar self-sustained oscillations may occur for a passive suspension that has low damping. Whereas the existence of this phenomenon was recognized as early as 1954 from incidental experimental results,[2] it was not understood until analyzed by Parente.[18,19] In the early experiments, self-sustained radial oscillations in the 10- to 100-hertz range were observed with an excitation frequency of 1000 hertz.

In Parente's analysis, which was made for a Case (1) magnetic suspension, Chapter 2, the lower bound to the viscous mechanical damping is given which determines whether the suspension is stable. If the damping is less than the given bound, the suspension displays force-free or self-sustained oscillations. For the oscillations to be force free, the angular frequency ω_e of the electrical supply and the mechanical angular frequency ω_m must be commensurate:

$$2\omega_e = k\omega_m,$$

in which k is an integer. The mechanical oscillation may be nonsinusoidal, but if the fundamental dominates,

$$\omega_e - \omega_m = \omega_d \approx \frac{1}{\sqrt{2LC}} = \omega_0,$$

in which ω_d is the damped natural angular frequency of the electric circuits with the suspended member centered, and ω_0 is the undamped natural frequency. For coils having high Q_L, these frequencies are nearly equal. In this notation, $2L$ is the self-inductance of a coil pair. Then

$$k = \frac{2\omega_e}{\omega_e - \omega_d} \approx \frac{2\omega_e}{\omega_e - \omega_0},$$

$$\omega_e = \frac{\omega_d k}{k - 2} \approx \frac{\omega_0 k}{k - 2},$$

and

$$\omega_m = \frac{2\omega_d}{k - 2} \approx \frac{2\omega_0}{k - 2}.$$

Similar relations are derived for the third harmonic in the motion dominant, and both actions have been well confirmed experimentally. Evidently, the mechanical motion can have a much lower frequency than the electric supply.

A simple partial explanation of the self-sustained oscillations of the passive suspension is as follows. A static curve of coil current versus coil self-inductance as the gaps of Figure 2-1 change is given in Figure 2-2, and the static curves of force versus displacement are shown in Figures 2-3 through 2-5. The static force-displacement curves really are symmetrical about the origin, so that the curves for negative displacements would be symmetrical to the curves for positive displacement but located in the second quadrant. When the gaps are changing rapidly, the dynamic force-displacement curves may differ considerably from the static force-displacement curves, depending on the electric-circuit parameters, because for a certain displacement and gap length and corresponding inductances, the currents and hence the corresponding

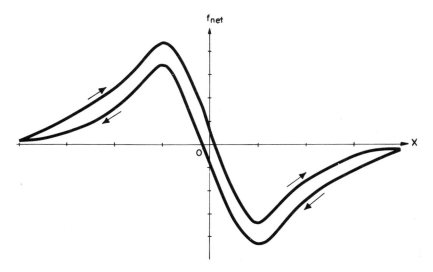

Figure 11-30 Dynamic force-displacement curve for passive suspension in oscillation.

forces are either larger or smaller than the corresponding static currents and forces. When the gap is closing and the inductance increasing, the current decay is delayed, and point for point the force is larger than the static force. When the gap is opening and the inductance decreasing, the current buildup is delayed, and point for point the force is smaller than the static force. The net result is illustrated in Figure 11-30. On the average over a cycle, the electromagnetic force does work on the moving body and therefore is able to sustain its motion against friction and windage or other restraining forces, and possibly supply useful work.

From the standpoint of application to floated instruments, all these oscillatory possibilities are undesirable, but they are mostly rendered harmless by the heavy damping used. However if conditions permit low-frequency self-sustained oscillations of a frequency near the forced-oscillation frequency caused by an unbalanced gyro wheel, the situation could be difficult. On the other hand, the self-sustained oscillations of the passive suspension system may be directed to useful purposes, as explained in the next section.

11-8 Potential Applications and Spin-offs

Most devices or ideas originally developed for rather specific applications eventually turn out to have various other uses, some of which may be more

important or more successful than the first intent. In fact, the original scheme sometimes proves to be a rather complete failure, and a quite unanticipated application becomes a spectacular success. However, in this instance, the original intent of the magnetic suspension for use in improving the performance of inertial instruments for sophisticated guidance and navigation has had spectacular success. But the idea has a number of other potential applications and spin-off uses, and the applications in guidance and navigation still have possibility for further exploitation in development and refinement.

The GDM Viscometer,[89] invented by Gilinson, Dauwalter, and Merrill,* for the determination of viscosities of blood and plasma, came into being through an inquiry directed to the Draper Laboratory by Professor Merrill about the possibility of using a magnetic suspension for such a device. Though the original intent was to use a magnetic suspension, and such a development still may be pursued in hope of increasing accuracy of measurement by another order of magnitude, lack of developmental funds at the time led to the use of an already existing air bearing. The use of this instrument by Professor Merrill, in collaboration with doctors of the Harvard Medical School, has resulted in important discoveries relating to the clotting properties of blood and other blood structure phenomena.[63, 90]

An improved float can is under development that is expected to have better accuracy than the vertically balanced pivot apparatus described in Section 10-2 and to be more convenient than the air-bearing apparatus described in that section. This proposed apparatus is illustrated in Figure 11-31. Instead of having a pivot, the apparatus for supporting and locating the member under test is to be an active magnetic suspension of the modern Gilinson-Scoppettuolo type or the Scoppettuolo time-sharing type. The pair of tapered Ducosyns are to provide radial and axial centering by means of the direct-current force windings under the control of force-to-balance loops taken from alternating-current radial and axial position signals, as shown in Figure 11-31 for the control of torque about the z axis by the torque-to-balance loop taken from the Microsyn signal generator part of the Ducosyn at the top to the Microsyn torquer part of the Ducosyn at the bottom. To take radial or axial force-displacement curves on a suspension under test, small signal voltages may be inserted in the appropriate force-to-balance loops of the test fixture to produce the desired deflections. The direct force currents may be calibrated in terms of force, and the displacement signals may be calibrated in terms of displacement. For radial displacements the bottom Ducosyn may be

* Professor of chemical engineering, Massachusetts Institute of Technology.

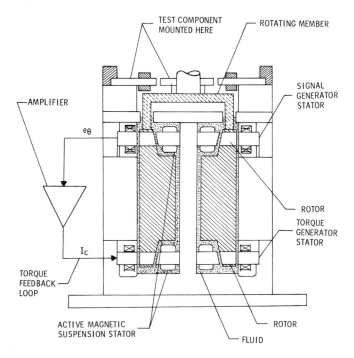

Figure 11-31 Proposed improved float can.

held fixed, if desired, to act essentially as a pivot, or parallel displacements may be produced by motions of both Ducosyns, in which case a force summation is necessary. Likewise for axial displacements a force summation is necessary. These summations are to be accomplished automatically in calibrated indicating instruments on a control panel. The distance between signal center and force center may be determined by reading the radial displacement signals from the test fixture with the suspension under test unenergized, and again with it energized, and taking the difference. This displacement is a measure of the side loading. The current supplied to the Microsyn torquer in the test fixture will be a measure of the spurious torque of the suspension under test and may be read directly from a calibrated instrument.

An oscillating viscometer using a principle similar to dynamic centering has been invented by Gilinson and Dauwalter;[91] it presently uses an air bearing but may be adapted to use apparatus similar to the magnetic float can just described. Voltages signaling torque commands may be inserted in

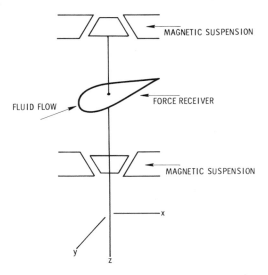

Figure 11-32 Schematic for proposed fluid-flow meter.

a feedback loop, just as voltages signaling force commands are inserted in the feedback loop for radial control in dynamic centering, as described in Section 10-3. The present instrument has been used quite successfully in measuring the viscosity of blood and plasma and various sinovial fluids. It is used also in obtaining torque-angle curves in the production testing of single-degree-of-freedom floated instruments.

The magnetic suspension has been proposed by Oberbeck as the sensing element in a highly accurate fluid-flow meter. A possible construction is shown schematically in Figure 11-32, in which a thin-wall "float," to be made of material of density close to the density of the fluid in which it is to be immersed, is suspended along three axes by magnetic suspensions. The float is to be adjusted so that at a reference temperature the buoyancy is essentially equal to the weight, and may be streamlined to orient itself in the direction of flow. For single-axis flow, such as through a pipe or trough, the x axis would be mounted in the flow direction, which must be horizontal. The y axis would be mounted vertically, and the z axis would be mounted horizontally, normal to the flow direction. The force exerted on the float in the x direction would be proportional to the flow velocity $\dot{x}_f$, and the force exerted on the float in the y direction would be proportional to the change in density of the fluid, or proportional to the decrease in temperature $-\Delta T_f$ of the fluid from reference temperature. For a passive suspension, the displacement is essentially proportional to the applied force, for small displacements, so that the displacements signals for the x axis can be taken as measures of flow, and the displacement signals for the y axis can be taken as measures of temperature change:

$$\dot{x}_f = k_f E_{Bx}$$

and

$$\Delta T_f = -k_T E_{By}.$$

The z-axis suspension force would serve merely to keep the float centered. This device might be useful in connection with water meters, gasoline pumps, or flow of brewery or winery products. For a two-axis flow region, the x axis and the y axis would be mounted in the plane of flow, which must be horizontal, and the z axis would be vertical. Then

$$\dot{x}_f = k_f E_{Bx},$$

$$\dot{y}_f = k_f E_{By},$$

and

$$\Delta T_f = -k_T E_{Bz}.$$

This device might be useful in connection with mixing valves or in other confluences, study of tides, ocean currents, or currents in other large bodies of water. For three-axis flow, the float would have to be spherical. The signals from the respective axes would be measures of the flow velocities along those axes, except that a separate temperature measurement would be needed to correct the z-axis (vertical) flow signal. This device probably would be better suited than the two-axis device for study of tides and currents. Since the electrical displacement signals are alternating voltages, the instantaneous flow rates or temperature changes can be indicated by voltmeter readings or by recorder charts, and the integrated flow can be obtained by use of watthour meters the current coils of which carry fixed alternating reference currents. For two- or three-dimensional flow, the components may be integrated separately, or the total may be obtained by using a polyphase meter. Also the intake and discharge of heat-exchanger water might be governed and equalized by feeding both the intake and discharge flow signals to the same integrating meter.

GRAVITY GRADIOMETER BASED ON 2FBG 10H GYRO DESIGN
(LESS THERMAL JACKET)

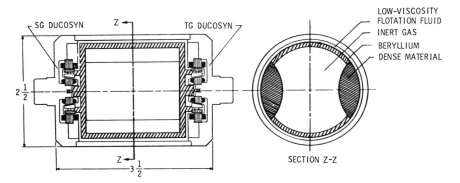

Figure 11-33 Gradiometer with Ducosyn mounting.

A gradiometer system[92,93] for gravity anomaly surveying is under development by Trageser* for use in a moving vehicle. A performance level in gravity gradient of $\frac{1}{3} \times 10^{-9}$ Eötvös unit ($\frac{1}{3} \times 10^{-9}$ sec^{-2}) is projected, which gives a survey accuracy of $\frac{1}{2}$ milligal (0.001 cm sec^{-2}) over distances of 10 miles. Various instrument configurations are described in the cited references and may be floated and suspended like gyros or accelerometers and mounted on a stabilized platform. In Figure 11-33 is shown an arrangement similar to a modern integrating gyro using Ducosyns, but the float instead of housing a gyro is merely a shell of beryllium in which are mounted very dense masses to sense the pull of gravity. However, use of this system is undesirable due to the considerably different demands that are made on the gradiometer system compared with the gyro system. In particular, in the gradiometer system the torques and the centering forces required are very much less than required for the gyro system, and the tolerable uncertainty torques of the gradiometer are very small. Excessive forces and torques as well as thermal gradients in the flotation fluid contribute to the uncertainty torques. These circumstances have dictated the use of capacitive signal generators and torquers and an active electric suspension. The circuits are pulse-operated and multiplexed, so that part of the time the central circuits on the float and its

* Milton B. Trageser, an associate director, Draper Laboratory.

housing operate as rotational position signal generators, part of the time they operate as torquers, and part of the time the capacitor plates of the suspensions for radial and axial centering are pulsed in sequence. The temperature of the jacket wall is to be controlled within 20 microdegrees Celsius. Thus the torque and force levels should be adequate, and the power requirements with attendant losses and heating should be very low. The parts of the device are shown in Figures 11-34, 11-35, and 11-36 prior to assembly. The device promises to have many uses. It may be a vertical deflection indicator for inertial navigation, the limit of accuracy of which is determined by the exactness with which the vertical can be known. It should be capable of producing gravity anomaly data of essentially the same quality as currently obtained by fixed-site free-air gravimeter observations and leveling. It may be used for inspection of one satellite by another. It may be used for topographical surveying from airplanes, for detection of tunnels or buried or submerged objects, for terrain avoidance by submarines, or for satellite weighing.

Magnetic or electric suspensions have been proposed for a large telescope to be mounted in a space vehicle. Though the telescope might be several feet in diameter and around 10 feet long, in weightless space the suspensions would merely be required to nudge it to keep it aligned with the vehicle. An electric suspension would have the advantage of requiring neither large masses of magnetic material attached to the telescope body or forming the stator frame, nor large coils that could require substantial power. Further, if the electric suspension were made active, the stiffness might be made substantially larger and the power requirements substantially smaller than would be possible for a passive suspension.

Magnetic suspensions already have been designed [94] for a 1-meter-diameter telescope on the Orbiting Astronomical Observatory (OAO) with the objective of obtaining 0.01-arc-second stability. The Skylab solar telescope [95] experiment uses flexural pivots for stability, but comparison between this scheme and the use of magnetic suspensions on the OAO telescope [96] are not available at this time. In the late 1970s an experiment [97] that includes a 3-meter-diameter telescope demanding a 0.005-arc-second stability is planned.

A telescope in a space vehicle may serve not only for astronomical observations but to keep the vehicle correctly oriented, or *pointed*. In the projected Manned Space Station [96] for the 1970s, the pointing stability for maintaining the station in a zero-gravity orbit will be most important. In addition to the telescope, the isolation of various experimental instruments from vibration

Figure 11-34 Parts of gradiometer float using capacitive signal and torque generators and electric suspension, featuring rotor. The scale shows one inch.

will be very important, a function that can be served by magnetic or electric suspensions. Further, magnetic and electric suspensions may play an important role in connection with the unmanned Fly-by and Swing-by missions to the outer planets.

The two-axis force generator, or *Wobulator*, developed by Oberbeck, in effect uses the magnetic suspension idea not to give a reaction force but to generate a force in the x-y plane whose direction and magnitude can be controlled electromagnetically. In Figure 11-37 a time-invariant flux density component $\mathscr{B}$ is obtained at each gap by means of direct-current excitation or by means of permanent magnets, superposed on which is an alternating flux density of amplitude $\mathscr{B}_m$:

$$\mathscr{B}_{81} = \mathscr{B} + \mathscr{B}_m \cos \omega t,$$

$$\mathscr{B}_{23} = \mathscr{B} + \mathscr{B}_m \sin \omega t,$$

$$\mathscr{B}_{45} = \mathscr{B} - \mathscr{B}_m \cos \omega t,$$

and

$$\mathscr{B}_{67} = \mathscr{B} - \mathscr{B}_m \sin \omega t.$$

Figure 11-35 Parts of gradiometer case using capacitive signal and torque generators and electric suspension, featuring stator. The scale shows one inch.

Figure 11-36 Exploded view of gradiometer. The scale shows one inch.

Since the pull at each pole is proportional to the square of the flux density there, the resultant pull is a vector rotating at angular velocity ω:

$$f(\mathcal{B}, \mathcal{B}_m, \omega) \propto 4\mathcal{B}\mathcal{B}_m e^{j\omega t}.$$

Alternatively, the excitation may be from rectified alternating current as shown in Figure 11-38, which gives a rotating force vector

$$f(\mathcal{B}, \omega) \propto \mathcal{B}^2 e^{j\omega t}.$$

Whereas the force vector rotates, the "rotor" does not; it *wobulates*. A principal application is for an *x-y* optical deflector, wherein a mirror is tilted through small angles about two orthogonal axes, under control of electrical signals, for oscillographic recording of data on film or for tracking of images such as stars. In this application, a mirror is mounted on the "rotor" by means of a flexible cylindrical structure that is provided on its axis near the center, Figure 11-39, so that the mirror can be caused to wobulate about the pivot. Other possible applications are for a slow mechanical scan television or for positioning of diffraction crystals in experimental radiation environments.

A three-axis spherical accelerometer using the six-pole magnetic suspension of Figure 7-20(b) was built by Gilinson about 1955. It had a thick spherical shell of Hipernik floated in a fluorocarbon damping fluid. The device was a failure largely because of the eddy currents induced in the sphere. Further, the radial displacement signals, which are supposed to be measures of the accelerations along the respective axes, are valid only for very small displacements, owing to the magnetic coupling that exists among the windings for the three axes and influences the signals. Whereas use of a ferrite ball later

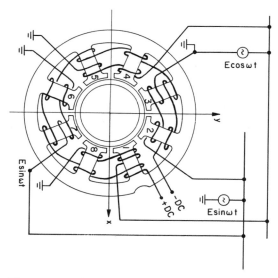

Figure 11-37 Schematic for Wobulator, using direct current with alternating current superposed.

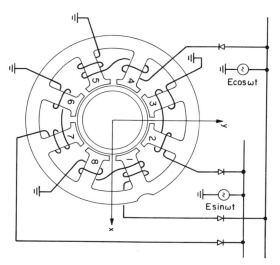

Figure 11-38 Schematic for Wobulator with direct-current excitation derived through rectification.

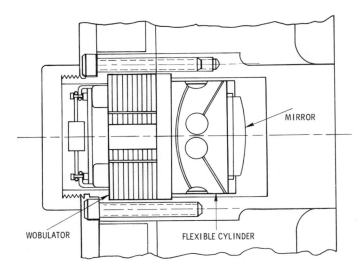

Figure 11-39 Wobulator applied to x-y optical deflector.

eliminated the eddy-current difficulty and made the device usable, the mag-
netic-coupling problem still presents a limitation. The coupling problem can
be eliminated by using the structure of Figure 7-20(a), and thus achieve a
simple device of considerable potential promise. Such a device might have
use for investigating dynamic processes in the fields of oceanography and
ocean engineering or meteorology. The instrument might replace the old
turn-and-bank indicators in aircraft. The three-axis spherical accelerometer
may be an adaptation of either the passive or the active magnetic suspension.

The trend toward the use of active suspensions has raised the possibility
of deriving a two-degree-of-freedom integrating gyro from a single-degree-
of-freedom instrument, especially for emergency use if one of three mutually
orthogonal instruments in an inertial navigations system should fail. In this
application,[98, 99] suggested by Fertig, the radial displacement signals of a
magnetic suspension would be used for detection of rotation of the gyro float
about the input axis IA, Figure 11-13, for torques applied about the output
axis OA. During the ill-fated Apollo 13 moon mission of April 1970, when
most of the command module's fuel cells were rendered inoperative, in order
to conserve the relatively small energy supply left in the lunar module, all but
the most vital life support equipment was subjected to "power-down," which

included lowering the heater power to the isolation platform gyros and accelerometers of the inertial guidance system. This situation allowed the damping fluid to cool and begin to congeal, and the temperature gradients in it could have caused the instruments to move from their proper orientations. Shortly before reentry into the earth's atmosphere, the apparatus was energized. Since the ship was still traveling in a zero-gravity environment, no large inertial forces were acting on the instruments during the power-down interval, but the substantially increased viscosity of the damping fluid nevertheless placed an extra demand on the suspensions for centering, which apparently was successfully met, as witnessed by the near perfect splashdown position. However, the additional security that might have been provided by active suspensions could have been very desirable in this situation and could be desirable in other situations. Incidentally, the radial and axial displacement signals from the magnetic suspensions in the navigation or guidance systems of space vehicles and missiles can be monitored on the earth at distances of thousands or even millions of miles. In fact, the analyses to determine whether the suspensions could function properly if the heaters were de-energized after the Apollo 13 accident were made at a ground station.

The self-sustained oscillations discovered in connection with the testing of passive magnetic suspensions can be utilized to operate reciprocating devices such as small pumps, compressors, or fans at low speeds without the use of gearing or mechanical linkages or other means of speed reduction directly from alternating-current supply lines of 60 hertz or higher frequency. Though the efficiency is quite low, for small devices for which energy economy is not an objective, the disadvantage is offset by simplicity of construction and operation and by the fact that no switching devices, either mechanical or electronic, are needed. Therefore the drive design can be rugged for long life without servicing or repairs. For a fixed electrical frequency, the mechanical frequency can be controlled by adjustment of the electric-circuit parameters, and a particular frequency can be encouraged by introduction of a spring and mechanical tuning. A small "forever fan" which utilizes the self-oscillation principle is under development at the Emerson Electric Company under the direction of L. W. Wightman.*

Magnetic suspensions may be used in the stabilizing devices for proposed high-speed rail cars. In fact one form of magnetic suspension has been used successfully in Germany for this purpose.

---

* Vice-president, manufacturing, research, and development.

11-9 Future Needs for Deep Space Navigation and Other Prolonged Uses

For journeys into deep space, a magnetic suspension and other components of a guidance system might be in a state of dormancy for months and then need to be placed into service without undue delay. With the spacecraft quietly cruising in zero-acceleration environment, the position need not be continuously monitored, and the entire magnetic suspension system may be shut down to economize energy. Periodically a small excitation signal could permit all the radial and axial signals to be searched, perhaps every 30 minutes. If no signals have exceeded prescribed thresholds, the system would be permitted to be dormant for another 30 minutes, and so on. But if any drift limits are exceeded, especially if they have been exceeded substantially for some unexpected reason, or if a heater should fail, prompt repositioning may be highly desirable.

For rapid restoration of position an active magnetic suspension is superior to a passive magnetic suspension of comparable size, and for restoration of z-axis position separate axial suspensions are desirable. Further, the active suspension requires less energy than the passive suspension. For the ultimate in energy economy, the active electric suspension is theoretically the answer, but at a sacrifice of force level and with possible complications due to grounding problems.

Similar dormant periods may exist when missiles are stored in a semiready state in underground silos, in long-range guardian or surveillance submarines that might lie on the ocean bottom for 30-day periods, for observer orbital satellites or communications orbital satellites, and for the orbiting terrestial space station and associated space shuttle. In all of these applications need for very stiff high-force suspensions and need for energy economy point to the active magnetic or electric suspension.

At the time of the development of the Leis suspension in 1965, the specifications were that its null torque should not exceed 0.1 dyne-centimeter and that the spurious restraint torque should not exceed 0.05 dyne-centimeter per milliradian of angular displacement. In unmanned deep space missions, such as the proposed "grand tour," these tolerances may need to be considerably tightened. Likewise for manned missions, such as Apollo, if the program continues, new navigation and guidance equipment is likely to be needed, either improvement of existing designs or entirely new designs. Today, for deep-space guidance equipment, the goals are for spurious torques not to exceed 0.001 dyne-centimeter and for angular position uncertainties not to exceed 0.001 arc-second.

11-10 Summary

Owing to the demand for higher and higher accuracies, especially in space navigation, and the need for energy economy, improved methods of testing and adjusting suspensions are under development, especially with respect to centering. Active magnetic and electric suspensions are under development to achieve reliability equal to or even better than the reliability of passive suspensions and supersede them, but the reliable and rugged passive magnetic suspension continues to be used and is expected to continue to have much use in the future. Though the development of magnetic and electric suspensions at the Draper Laboratory has been aimed primarily toward applications in the navigation and control of vehicles on land and sea and in the air and outer space, numerous other direct and spin-off applications have evolved, and numerous others are likely to follow.

REFERENCES

1. P. J. Gilinson, Jr., and J. Scoppettuolo, "A Microsyn Type of Magnetic Suspension for a Single-Degree-of-Freedom Torque Summing Member," Report E-240, Cambridge, Mass.: M.I.T. Instrumentation Laboratory, January 1953.

2. P. J. Gilinson, Jr., R. G. Haltmaier, G. A. Oberbeck, and J. A. Scoppettuolo, "A Single Winding Microsyn Magnetic Suspension," Report E-250, Cambridge, Mass.: M.I.T. Instrumentation Laboratory, March 1953.

3. P. J. Gilinson, Jr., "A Star-Connected Microsyn Magnetic Suspension," Report E-313, Cambridge, Mass.: M.I.T. Instrumentation Laboratory, July 17, 1953.

4. P. J. Gilinson, Jr., "An Alternating Current Microsyn Magnetic Suspension," Report E-395, Cambridge, Mass.: M.I.T. Instrumentation Laboratory, August 18, 1954.

5. P. J. Gilinson, Jr., "An Alternating Current Microsyn Magnetic Suspension," Report E-522, Cambridge, Mass.: M.I.T. Instrumentation Laboratory, February 1956.

6. P. J. Gilinson, Jr., G. E. Garcia, and J. H. Aronson, "8-Pole Microsyn Magnetic Suspension," Report E-588, Cambridge, Mass.: M.I.T. Instrumentation Laboratory, October 5, 1956.

7. P. J. Gilinson, Jr., "Microsyn Magnetic Suspension Elementary Principles," Report E-597, Cambridge, Mass.: M.I.T. Instrumentation Laboratory, November 1956.

8. P. J. Gilinson, Jr., "Supports for Rotating or Oscillating Members," U.S. Patent No. 3,184,271, issued May 18, 1965, filed May 15, 1957.

9. J. Hovorka, J. W. Hursh, E. J. Frey, W. G. Denhard, L. R. Grohe, P. J. Gilinson, Jr., and W. Vander Velde, "Recent Progress in Inertial Guidance," Am. Rocket Soc. J., vol. 29, pp. 946–957, 1959.

10. C. S. Draper, "The Inertial Gyro—An Example of Basic and Applied Research," Am. Sci., vol. 48, pp. 9–19, March 1960.

11. P. J. Gilinson, Jr., W. G. Denhard, and R. H. Frazier, "A Magnetic Support for Floated Inertial Instruments," Report R-277, Cambridge, Mass.: M.I.T. Instrumentation Laboratory, April 1960. Also appeared as Sherman M. Fairchild Fund Paper FF-27, Institute of the Aeronautical Sciences, New York, May 1960. A condensation appeared in Proceedings of the National Specialists Meeting on Guidance of Aerospace Vehicles, pp. 56–83, Boston, May 1960.

12. C. S. Draper, W. Wrigley, and J. Hovorka, Inertial Guidance, New York: Pergamon Press, 1960.

13. N. P. Chironis, "just around the corner? Magnetically Suspended Bearings," Prod. Eng., pp. 46–47, July 24, 1961.

14. S. Lees, ed., Air, Space and Instruments (Draper Anniversary Volume), New York: McGraw-Hill Book Company, 1963. Especially chapter on "Electromechanical Components" by P. J. Gilinson, Jr., and R. H. Frazier, pp. 285–350.

15. R. H. Frazier, "Signal and Torque Errors and Centering Forces of Microsyns and Magnetic Suspensions Due to Eccentricities of Rotor Position," Report E-1045, Cambridge, Mass.: M.I.T. Instrumentation Laboratory, November 1964.

16. C. C. Perez, "A Feedback System for a Stable Magnetic Suspension in Air," S.M. Thesis, Department of Electrical Engineering, M.I.T., June 1962.

17. R. H. Miller, "Synthesis of Electromagnetic Suspension Circuits," Ph.D. Thesis, Department of Aeronautics and Astronautics, M.I.T., January 1964.

18. R. B. Parente, "A Functional Analysis of Systems Characterized by Nonlinear Differential

Equations," Ph.D. Thesis, Department of Electrical Engineering, M.I.T., August 1965, chap. 7, "An Analysis of a Magnetic Suspension Device," pp. 135–167, and app. A, pp. 171–184.

19. R. B. Parente, "Stability of a Magnetic Suspension Device," *IEEE Trans.*, vol. AES-5, pp. 474–485, 1969.

20. J. E. Chrisinger, "An Investigation of the Engineering Aspects of a Wind Tunnel Magnetic Suspension System," S.M. and A.E. Thesis, Department of Aeronautics and Astronautics, M.I.T., June 1959.

21. E. L. Tilton III, W. J. Parkin, E. E. Covert, J. G. Coffin, and J. E. Chrisinger, "The Design and Initial Operation of a Magnetic Model Suspension and Force Measurement System," Technical Report 22, Cambridge, Mass.: M.I.T. Aerophysics Laboratory, August 1962.

22. "Magnetic Bearings for Aerospace Applications," Technical Documentary Report No. 63–474, Schenectady, N.Y.: Advanced Technology Laboratories, General Electric Company, May 1963.

23. J. D. McHugh, "Possibilities and Problems of Magnetic and Electrostatic Bearings," *Des. Abstr.*, p. 191, November 5, 1964.

24. "Survey of Magnetic Bearings," Cambion Latest Developments, Cambridge, Mass.: Cambridge Thermionic Corporation, 1965, 1972.

25. "Manual of Watthour Meters," GET-1840B, Somersworth, N.H.: General Electric Company Meter Department, 1952, p. 14.

26. D. F. Wright, "The Design of a Repulsion Magnetic Bearing for Watthour Meters," *AIEE Trans.*, vol. 80, pt. 3, *Power Apparatus and Systems*, pp. 755–758, 1961.

27. F. T. Holmes, "Axial Magnetic Suspensions," *Rev. Sci. Instrum.*, vol. 8, pp. 444–447, 1937.

28. L. E. MacHattie, "The Production of High Rotational Speed," *Rev. Sci. Instrum.*, vol. 12, pp. 429–435, 1941.

29. J. W. Beams, J. L. Young III, and J. W. Moore, "The Production of High Centrifugal Fields," *J. Appl. Phys.*, vol. 17, pp. 886–890, 1946.

30. J. W. Beams, "Magnetic Suspension Ultracentrifuge," *Electronics*, vol. 27, no. 3, pp. 152–155, 1954.

31. J. B. Breazeale, C. G. McIlwraith, and E. N. Dacus, "Factors Limiting a Magnetic Suspension System," *J. Appl. Phys.*, vol. 29, pp. 414–415, 1958.

32. A. W. Jenkins, Jr., and H. M. Parker, "An Electromagnetic Support Arrangement with Three Dimensional Control, Part I—Theoretical," Report UVA/ORL-04-58 TR1, Charlottesville, Va.: University of Virginia, 1958.

33. H. S. Fosque and G. H. Miller, "An Electromagnetic Support Arrangement with Three Dimensional Control, Part II—Experimental," Report UVA/ORL-04-58 TR3, Charlottesville, Va.: University of Virginia, 1958.

34. H. Sixmith, "Electromagnetic Bearing," *Rev. Sci. Instrum.*, vol. 32, pp. 1196–1197, 1961.

35. P. J. Geary, *Magnetic and Electric Suspensions* (a bibliography), SIRA Survey of Instrument Parts Series, vol. 6, South Hill, Chislehurst, Kent, England: British Scientific Instrument Research Association, 1964.

36. H. C. Roters, *Electromagnetic Devices*, New York: John Wiley & Sons, 1941, chap. 8.

37. G. A. Oberbeck, "Energy Storage Components Used as Signal or Force Devices," Report E-1565, Cambridge, Mass.: M.I.T. Instrumentation Laboratory, May 1964.

38. R. H. Frazier, "Magnetic and Electric Suspensions with Coil and Capacitor in Parallel," Report E-2349, Cambridge, Mass.: M.I.T. Instrumentation Laboratory, October 1968.

39. P. J. Gilinson, Jr., and R. H. Frazier, "Bridge-Circuit Connections for Passive Magnetic and Electric Suspensions," Report E-2737, Cambridge, Mass.: M.I.T. Charles Stark Draper Laboratory, February 1973.

40. J. G. Flick, "Performance of a Two Dimensional Four-Pole Square-Wave Driven Magnetic Suspension," S.B. Thesis, Department of Electrical Engineering, M.I.T., June 1966.

41. L. Y. Chin, "Analysis of a Floated Electrostatic Pendulum," S.M. Thesis, Department of Aeronautics and Astronautics, M.I.T., December 1967.

42. R. H. Frazier, "Combined Magnetic and Electric Suspension," Report E-1988, Cambridge, Mass.: M.I.T. Instrumentation Laboratory, July 1966.

43. R. H. Frazier, "Analysis of Four-Pole Magnetic Suspension for Cylindrical Core," Report E-743, Cambridge, Mass.: M.I.T. Instrumentation Laboratory, April 1958.

44. R. H. Frazier, "Maximum Stiffness of Magnetic and Electric Suspensions," Report E-848, Cambridge, Mass.: M.I.T. Instrumentation Laboratory, September 1959.

45. R. H. Frazier, "Eight-Pole Magnetic Suspension with Series-Parallel Connections," Report E-2350, Cambridge, Mass.: M.I.T. Instrumentation Laboratory, October 1968.

46. P. J. Gilinson, Jr., and R. H. Frazier, "Mesh Connections for Passive Magnetic Suspensions," Report E-2752, Cambridge, Mass.: M.I.T. Charles Stark Draper Laboratory, March 1973.

47. R. H. Frazier, "A Six-Pole Cylindrical Magnetic Suspension," Report E-1599, Cambridge, Mass.: M.I.T. Instrumentation Laboratory, June 1964.

48. R. H. Frazier, "Analysis of Four-Pole Electric Suspension," Report E-746, Cambridge, Mass.: M.I.T. Instrumentation Laboratory, July 1958.

49. J. Feldman, "Electrostatic Signal and Support Test Viscous Integrator EVI # 695," Report GL-274, Cambridge, Mass.: M.I.T. Instrumentation Laboratory, June 29, 1962.

50. R. M. Bozorth, *Ferromagnetism*, New York: D. Van Nostrand Company, 1951.

51. W. V. Lyon, "Heat Losses in the Conductors of Alternating Current Machines," *AIEE Trans.*, vol. 38, pp. 1361–1395, figs. 3 and 4, pp. 1373 and 1374, 1919.

52. R. H. Frazier and C. Kingsley, Jr., "Notes for Electromechanical Components and Systems 6.42," 2d ed., Cambridge, Mass.: M.I.T., 1964, sec. 5-2 and sec. 9-4.

53. R. H. Frazier, "Initial Stiffness of Four-Pole Magnetic Suspension with Constant Effective Q_e," Report E-876, Cambridge, Mass.: M.I.T. Instrumentation Laboratory, December 1959.

54. R. H. Frazier, "Maximum Initial Stiffness of Magnetic Suspension with Constant Effective Q_e," Report E-902, Cambridge, Mass.: M.I.T. Instrumentation Laboratory, April 1960.

55. R. H. Wilkinson, "Float Motion Torques in a Floated Single-Degree-of-Freedom Integrating Gyro," Report E-2488, Cambridge, Mass.: M.I.T. Charles Stark Draper Laboratory, February 1970.

56. C. J. Albers, Jr., "Disaccommodation in Ferromagnetic Materials," S.B. Thesis, Department of Electrical Engineering, M.I.T., June 1965.

57. G. J. Caporaso, "A Mechanism for Disaccommodation in Mn-Zn Ferrites," S.B. Thesis, Department of Physics, M.I.T., June 1969.

58. "Electronic Amplifiers" (a compilation), NASA SP-5947(01). For sale by National Technical Information Service, Springfield, Va. 22151.

59. J. A. Scoppettuolo, "A Time-Sharing Active Analog Suspension," Report E-2501, Cambridge, Mass.: M.I.T. Charles Stark Draper Laboratory, June 1970.

60. M. D. Leis, "A Pulse Restrained Magnetic Suspension," S.M. Thesis, Department of Electrical Engineering, M.I.T., January 1966.

61. P. T. Hirth, "A Time-Sharing Active Magnetic Suspension," S.M. Thesis, Department of Electrical Engineering, M.I.T., September 1968.

62. P. J. Gilinson, Jr., "Radial Positioning of the Eight-Pole Magnetic Suspension Rotor," Report E-589, Cambridge, Mass.: M.I.T. Instrumentation Laboratory, October 1956.

63. P. J. Gilinson, Jr., C. R. Dauwalter, and E. W. Merrill, "A Rotational Viscometer Using an AC Torque-to-Balance Loop and Air Bearing," *Trans. Soc. Rheol.*, vol. 7, pp. 319–331, 1963.

64. P. J. Gilinson, Jr., C. R. Dauwalter, and J. A. Scoppettuolo, "A Multirange Precision Torque Measuring Device," Report R-367, Cambridge, Mass.: M.I.T. Instrumentation Laboratory, July 1962.

65. P. J. Gilinson, Jr., "Torque Measuring Systems," Report E-1562, Cambridge, Mass.: M.I.T. Instrumentation Laboratory, May 1964.

66. F. D. McCarthy, "Analysis of Drift," Report E-2298, vol. 1, Cambridge, Mass.: M.I.T. Instrumentation Laboratory, July 1968.

67. W. G. Denhard, ed., *Inertial Component Testing: Philosophy and Methods*, AGARDograph 120, Maidenhead, England: Technivision, 1970.

68. J. M. Slater, *Inertial Guidance Sensors*, New York: Reinhold Publishing Corporation, 1964.

69. R. E. Marshall and P. J. Palmer, "Inertial Instrument Design Verification Tests for High-G Applications," Report E-1866, Cambridge, Mass.: M.I.T. Instrumentation Laboratory, October 1965. (Presented at International Symposium on Inertial Guidance Testing Techniques, Wilkinson, Germany, November 1965.)

70. A. Truncale, W. Koenigsberg, and R. Harris, "Spectral Density Measurements of Gyro Noise," Report E-2641, Cambridge, Mass.: M.I.T. Charles Stark Draper Laboratory, February 1972.

71. R. K. Mueller, "Microsyn Electromagnetic Components," Report E-224, Cambridge, Mass.: M.I.T. Instrumentation Laboratory, December 1952.

72. R. K. Mueller, "Dynamo Transformer," U.S. Patent No. 2,488,734, issued November 22, 1949, filed March 7, 1946.

73. G. E. Garcia, "Combined Magnetic Suspension and Rotary Magnetic Transformer," U.S. Patent No. 3,079,574, issued February 26, 1963, filed January 6, 1960.

74. W. Wrigley, W. M. Hollister, and W. G. Denhard, *Gyroscopic Theory, Design, and Instrumentation*, Cambridge, Mass.: M.I.T. Press, 1969.

75. R. R. Ragan, "General Analytical Study of Cross-Roll Correction and Vector Integration," Report E-566, Cambridge, Mass.: M.I.T. Instrumentation Laboratory, September 1956.

76. J. E. Miller, ed., *Space Navigation Guidance and Control*, AGARDograph 105, Maidenhead, England: Technivision, 1966.

77. C. S. Draper, "Gyroscopic Apparatus," U.S. Patent No. 2,752,790, issued July 3, 1956, filed August 2, 1951.

78. J. Kirk, "Progress Report No. 2, Project Atlas," Report E-444, Cambridge, Mass.: M.I.T. Instrumentation Laboratory, January 1955.

79. R. R. Ragan, G. W. Mayo, and J. Feldman, "Apollo Guidance and Navigation Flight System Reliability," Report R-569, Cambridge, Mass.: M.I.T. Instrumentation Laboratory, March 1967.

80. J. J. Jarosh, C. A. Haskell, and W. W. Dunnell, Jr., "Gyroscope Apparatus," U.S. Patent No. 2,752,791, issued July 3, 1956, filed February 9, 1951.

81. J. P. Gilmore and J. Feldman, "Gyroscope in Torque-to-Balance Strapdown Application," *AIAA J. Spacecraft Rockets*, vol. 7, pp. 1076–1082, September 1970.

82. C. S. Draper, W. Wrigley, and L. R. Grohe, "The Floating Integrating Gyro and Its Application to Geometrical Stabilization Problems on Moving Bases," Sherman M. Fairchild Fund Paper FF-13, New York: Institute of Aeronautical Sciences, January 1955.

83. C. S. Draper, W. G. Denhard, and E. J. Hall, "Effects of Zero-G Conditions on Gyro Unit Operation," Report R-401, Cambridge, Mass.: M.I.T. Instrumentation Laboratory, December 1962.

84. M. S. Sapuppo, "Product and Environmental Specifications, 16 Pendulous Integrating Gyro Accelerometer, Model J," Report E-1668, Cambridge, Mass.: M.I.T. Instrumentation Laboratory, November 1964.

85. R. B. Mark, "A Wave Resolver for Compact Gimbal Systems," S.M. Thesis, Department of Aeronautics and Astronautics, M.I.T., June 1960.

86. R. L. Anderson, "Tests on the 360-Pole Wafer Resolver," Report E-1296, Cambridge, Mass.: M.I.T. Instrumentation Laboratory, March 1963.

87. K. Fertig and J. F. Hendrickson, "Angular Resolver," U.S. Patent No. 3,284,795, issued November 8, 1966, filed March 31, 1964.

88. L. R. Grohe, E. J. Hall, M. S. Sapuppo, and A. E. Scoville, "The M.I.T. 25 Series Inertial Instruments," Report R-141, Cambridge, Mass.: M.I.T. Instrumentation Laboratory, February 1958.

89. P. J. Gilinson, Jr., C. R. Dauwalter, and E. W. Merrill, "Viscometer," U.S. Patent No. 3,343,405, issued September 26, 1967, filed February 28, 1962.

90. E. W. Merrill, W. G. Margetts, G. R. Cokelet, and E. R. Gilliland, "The Casson Equation and Rheology of Blood Near Zero Shear," *Proceedings of the Fourth International Congress on Rheology, Part IV, Symposium on Biorheology*, ed. by A. L. Copley, New York: John Wiley & Sons, 1965, pp. 135–143.

91. P. J. Gilinson, Jr., and C. R. Dauwalter, "An Oscillating Viscometer," patent applied for and pending, filed March 16, 1972.

92. M. B. Trageser, "A Gradiometer System for Gravity Anomaly Surveying," Report R-588, Cambridge, Mass.: M.I.T. Charles Stark Draper Laboratory, June 1970.

93. M. B. Trageser, "Gravity Gradiometer Program," Report C-3615, Cambridge, Mass.: M.I.T. Charles Stark Draper Laboratory, October 9, 1970.

94. W. B. Chubb, D. N. Schultz, and S. M. Seltzer, "Attitude Control and Precision Pointing of the Apollo Telescope Mount," AIAA Paper No. 67-553, August 1967.

95. D. J. Chiarappa, "Fine Pointing and Stability of Space Station Experiments," AIAA Paper No. 71-62, AIAA 9th Aerospace Sciences Meeting, New York, January 25–27, 1971.

96. "Candidate Experiment Program for the Manned Space Station," NASA/MSFC NHB-7150XX, September 15, 1969.

97. Apollo Staff, "Guidance and Navigation Requirements for Unmanned Fly-by and Swing-by Missions to the Outer Planets," vols. 1–4, Cambridge, Mass.: M.I.T. Charles Stark Draper Laboratory, October 1971.

98. K. Fertig, "Applications of Inertial Guidance Technology for Space Navigation," Report R-650, Cambridge, Mass.: M.I.T. Charles Stark Draper Laboratory, October 1969. (Presented at the International Navigation Conference, October 28–30, 1969, Hamburg, Germany.)

99. J. M. Dare, "Two Degree Information from a Single-Degree-of-Freedom Gyro," S.M. Thesis, Department of Aeronautics and Astronautics, M.I.T., June 1970.

INDEX